The Domestic Duck

The Domestic Duck

CHRIS AND MIKE ASHTON

THE CROWOOD PRESS

First published in 2001 by
The Crowood Press Ltd
Ramsbury, Marlborough
Wiltshire SN8 2HR

www.crowood.com

Paperback edition 2008

This impression 2013

British Library Cataloguing-in-Publication Data
A catalogue record for this book is available from the British Library.

ISBN 978 184797 050 3

All ducks and photographs are the authors', unless otherwise credited.
Diagrams by the authors and David Fisher, unless otherwise credited.

Typeset by Carolyn Griffiths, Cambridge

Printed and bound in India by Replika Press Pvt. Ltd.

Contents

Acknowledgements

Many people have contributed to the shared knowledge of the breeds recorded in this book. Some of this material has not been previously written down and much of the written source material from the nineteenth and early-twentieth century is now difficult to find. We therefore felt it useful to record what is known about domestic waterfowl breeds in Britain at this time.

Special thanks are due to John Hall whose extensive knowledge and life-long involvement with waterfowl has proven especially valuable in the sections on Call, Crested, Cayuga and Hook Bill Ducks.

Tom Bartlett, Vernon Jackson, Stephanie Mansell, Gilbert May, Anne Terrell, Fran Harrison and many other British Waterfowl Association members have also played a part in giving their views on particular breeds.

On the continent, Kenneth Broekman and Hans Ringnalda have given advice on the Dutch breeds, and Evelyne van Vliet, Marie Buttery and Holgar Heyken have translated material from Dutch and German.

On the technical side, F. M. Lancaster was kind enough to check over, and give advice on, the introduction to plumage colour genetics in the 'Designer Ducks' section. Victoria Roberts has given up her time to proof-read and her help is much appreciated.

Thanks are also due to *Poultry World* and Sacrewell Poultry Trust for access to old copies of *Poultry World*; also *The Feathered World* for access to magazines from the early twentieth century.

Introduction

For thousands of years ducks have been a source of interest for human beings, mainly for very practical reasons – food and warmth. Duck down from many species, including the famous Eider, has continued to provide very efficient insulation for clothes and bedding. Duck eggs are a good source of protein and in many cultures duck meat has long been a favourite food. Traditionally these were obtained by rifling nests or capturing the birds themselves. For nomadic peoples, raiding nests and trapping or shooting the ducks would have been common activities in many parts of the world. Only when settled communities evolved, around the development of cereal agriculture, were more permanent means of keeping ducks made possible.

Although most ducks are good fliers, it is very easy to 'clip their wings' and impose simple captivity and, because many species nest on the ground, obtaining hatching eggs would be fairly straightforward. What is amazing is that out of 147 living species of ducks, geese and swans, only four of them have provided the bulk of domestic waterfowl throughout the world. Two of these are geese: the Greylag (*Anser anser*) and the Swan Goose (*Anser cygnoides*). Two of them are ducks: the Muscovy (*Cairina moschata*) and the Northern Mallard (*Anas p. platyrhynchos*). It is the latter that appears to be the ancestor of all but one of the breeds of domestic duck. The Common Mallard is the same species as the big Rouen and Aylesbury as well as the little Black East Indian and the tiny Call Duck. This is all too evident if a Wild Mallard drake flies into the breeding pens and manages to fertilize whatever domestic ducks are on the premises. These birds are highly sexed and highly successful. This is one of the prime reasons for choosing the Mallard in the first place.

*Fig 1 The Muscovy duck (*Cairina moschata*), a perching duck from South America, now common around the world as a farmyard duck. This is a different species from the Mallard, which produced all of our other domestic ducks.*

'Mallards rank among the most successful of all avian species, and throughout their wide distribution (which encompasses most of the Northern Hemisphere) they occupy a tremendous variety of habitats,' writes Frank S. Todd in his *Waterfowl: Ducks, Geese and Swans of the World* (1979). He goes as far as to assert that the Mallard may well have been the first domesticated bird, pre-dating even the chicken. There are records of the Romans establishing a duckery in the first century using Teal or Mallard eggs. The Egyptians certainly kept Mallard several centuries before the birth of Christ and the Chinese were probably keeping them much earlier. One tell-tale piece of evidence lies in the curled tail feathers of the Mallard, a feature found in all but one of the domestic breeds. Charles Darwin asserts: 'In the great duck family one species alone – the male of *Anas boschas* [now *Anas p. platyrhynchos*] – has its four middle tail feathers curled upwardly'.

Another characteristic shared by domestic and wild surface-feeding ducks is the speculum, a patch of iridescence on the secondary flight feathers. It is common to both male and female ducks and tends to be accompanied by black and white bars on the secondaries and greater coverts. This feature is retained in many of the domestic breeds, although it is obscured in some cases by masking genes and various dilutions.

Ancestors

Whether all of the breeds, other than the Muscovy, are derived from the single species of Northern Mallard is a matter for speculation. Certainly, in the wild there are close relative species that will interbreed with the Mallard, just as the two geese species mentioned above will hybridize to produce viable breeds like the Steinbacher and some of the Russian geese. One of the closest to the Mallard is a subspecies, *Anas p. conboschas*, known as the Greenland Mallard. It is larger and lighter-coloured than its common relative, partly marine in its habitat and requires two years to reach sexual maturity. It looks very similar to the Trout Indian Runner, apart from the upright stance and the Runner's single year to maturity. Another close full species is the American Black Duck, *Anas rubripes*, which will certainly hybridize with the Mallard, according to Todd (1979). This is a contender for the ancestor of the Cayuga and the Black East Indian Ducks, both of which were developed within the breeding range of *Anas rubripes*.

The question remains, however: if so many domestic breeds have emerged from such a small family of wild ducks, most of which are very similar to each other, why are the domestics so different? Why are there massive ones and minute ones, black ones, brown ones, white ones, pied ones and Mallard-looking ones? The answer is likely to be 'natural variation' or mutation. Within any species there is a range of genetic material. It is even possible to get the odd white blackbird, but within the constraints and competition for survival of the natural habitat only the most conservative and viable will survive. It is theoretically possible for wild Mallards to generate large white Aylesbury look-alikes, yet what would their chances of survival be in the wild compared to their lightweight, camouflaged, high-flying siblings? Natural selection favours the most efficient adapters to the specific environment. Take away the constraints of the wild; protect the birds from predators and bad weather; give them buckets full of high-protein food and the Aylesbury will survive as well as its wild brothers or sisters – at least until cooking time.

The Mallard has the potential for tremendous variation in size and colour, yet it is humankind that has seen fit to exploit this to develop the multiplicity of domestic ducks. This may even be fairly quick to achieve. The Rev. Dixon described his attempts to breed wild ducks. The eggs were taken from a wild Mallard's nest and hatched under an ordinary duck.

> Until a month old, we 'cooped' the old duck, but left the youngsters free. They grew up invariably quite tame, and bred freely the next and following years. There was one drawback however. Although not admitted when grown to the society of tame ducks, they always, in two or three generations, betrayed prominent marks of deterioration; in fact, they became domesticated. The beautiful carriage of the Wild Mallard and his mate, as seen at the outset, changed gradually to the easy, well-to-do, comfortable deportment of a small Rouen, for they, at each generation, became much larger.

Their wings no longer crossed over the rump in a ready flight position, but dropped down at their sides like the larger domestic ducks.

Classic Ducks

During the course of this book we will refer constantly to two kinds of domestic duck: the 'classic' and the 'designer'. We make no apologies for such coinage: the terms refer to very different forms of development, as we will try to explain.

The 'classic' ducks are typically those featured in the earliest books of waterfowl standards and those still tending to win most prizes in major exhibitions. They include the Aylesbury, Cayuga, Pekin and Rouen, in the heavy breeds, as well as the Bali, Black East Indian, Call, Hook Bill, Indian Runner

Fig 2 The common tame duck (the Mallard), ancestor of all our domestic ducks except the Muscovy.

Fig 3 The Muscovy Duck, Anas moschata. *(Illustrations from Willughby, 1678; photocopy; by permission of the British Library)*

and the Crested Duck, in the lighter varieties. What is special about these birds is the nature of their evolution compared with the more deliberate engineering of the later 'designer' ducks. Classic ducks evolved in different geographical areas over what may have been long periods of time. They were developed in some cases for meat, like the white Aylesbury with its clean-looking carcass, large body and light-coloured meat. They were also developed as economical foragers, like the so-called Indian Runner that gleaned its food from the padi fields of Indonesia and was famed for its prodigious egg-laying. They were developed as Decoy Ducks, the alternative name for the Call Ducks, because of their portable size and loud quacking, which brought wild ducks down to the guns of wild-fowlers or into the mouths of great cage traps (*koys*). Each 'breed' of duck had a very special pool of genetic material. Until the nineteenth century it is very unlikely that the last three mentioned birds were ever 'live' in the same continent together, let alone the same collection. They were unique simply because they were not mixed.

Although based on the same ancestor, the birds evolved in different geographical areas because of the limitations of human travel. As long as it took months to journey to China by horse, camel or sailing ship, there was little chance that voyagers would bother carrying anything so trivial as a duck, apart from the odd dinner. It is only when transport became easier and the Western nations began to open up their empires all over the globe that an interest in exotic waterfowl really 'took off'. Until then the British slowly developed their Aylesbury; the French developed a parallel form in the Rouen; the Chinese produced the Pekin; and the archipelagos of South East Asia concentrated on the perambulatory egg-layer, the Indian Runner Duck. These bird populations were largely restricted to distinctive gene pools because of the original travel problems.

*Fig 4 Pekin drake. The property of Mr F. A. Miles: first prize Crystal Palace 1910, etc. (*The Feathered World, *23 June 1911)*

Whilst global contact was limited, the ancient duck populations of Indonesia and China remained relatively isolated from Western culture. Exotic specimens were a rarity and usually arrived back in Britain stuffed, skinned or preserved. However, during Victorian times there was a transport technology revolution, which accompanied the manufacturing revolution, so that journey times were shortened and it became more likely that livestock would survive long journeys back from the Far East. The Dutch and British had also established their colonies and trading posts in Indonesia, Malaya and India giving the West the contacts to bring these new breeds home. The American trade with China and Japan resulted in the USA importing live birds from the Far East, including Chinese and 'African' geese as well as the Pekin duck. The more affluent cultures of the West with their Victorian mania for collection and cataloguing were not only interested in the birds for their commercial potential; the new imports also caused much interest in their novelty and aesthetic appeal. So although the Pekin did revolutionize the duck industry in both Britain and America, the imported birds were also of interest at a show.

It was in Victorian times that the idea of the show or exhibition actually began. The show was for entertainment as well as a shop window for agricultural produce. Initially the show duck was just a carcass; the dressed duck was all that was on exhibit, amongst the goose and chicken carcasses in the poultry section. However, as the era progressed, so did the number of breeds. The Aylesbury competed with the Pekin in the commercial world and the Rouen was also developed as a heavy duck. The Indian Runner arrived as a novelty egg-layer, probably at considerable individual expense. Call Ducks and Black East Indians were noticed as new and unusual ducks. Why show them dead? These breeds had not only commercial but aesthetic appeal, and the idea of showing valuable, newly imported breeds alive took off after the first show of feathered exhibits at the Crystal Palace in 1845. Classes were provided for:

- Aylesbury or other white variety duck;
- any other variety duck;
- common geese;
- Asiatic or Knob Geese;
- any other variety geese.

The idea of the duck, goose and chicken as an exhibition bird and hobby began to replace their original image as a meal.

Designer Ducks

Almost as soon as the imported 'classic' ducks reached Europe they were mixed with local blood lines. Indian Runners in Cumbria were crossed with farmyard birds until their carriage was half-way between the vertical Runner and the horizontal Aylesbury. The Americans too crossed their own Aylesbury with newly imported Pekin; as a result the standard American Pekin is still much less upright than those coming from Germany. Although there was a danger of contaminating the purity of the imported blood lines, it was soon found that there were serious advantages. Completely new gene pools were merged, allowing greater variety of shape, size and colour. There was an influx of vigour common to many hybrids and an opportunity for people to create their very own breeds. The late nineteenth century opened the flood-gates of innovation and experimentation. Those breeders who could stabilize a pattern or colour were able to claim celebrity status, especially if they could make capital on the hybrid vigour that certainly gave a boost to the egg-laying production. Basically, the situation at the turn of the last century was one of playing with the new genes rather than genetic engineering in the modern sense.

There is also the chance to 'customize'. If you have a sufficient range of varieties, it is always possible to plan the next generation of waterfowl. By mixing an egg-layer with a table bird, then crossing it back to another variety, you can develop a multi-purpose variety or one with even more extreme capabilities in the table or egg department. By concentrating also on the colours of the birds you can also end up with something both unique and surprising.

It does not take a great leap of the imagination to appreciate the 'glory' attached to having a breed named after you. It is almost as good as having a country, Rhodesia, for example, or America, Bolivia or Colombia. The Victorians particularly loved that aspect of notoriety; that is partly why they scattered named grave-stones so liberally around the country-side, and even named their trout flies 'Greenwell's Glory' or 'Wickham's Fancy'. Create a special new breed of ducks and immortality is yours . . . at least if you are called Campbell, or live in Orpington or if your farm is called the Abacot Duck Ranch!

The ducks developed from the crossings and re-crossings of the new imports, especially the Indian Runners and the Pekins at the end of the nineteenth century, are what we have chosen to call the 'designer' ducks. You will notice that most of them have been classified in the show programmes as 'Light Ducks'. This tends to indicate an egg-layer or general purpose bird derived from the Runner. They include the Abacot Ranger, the Campbell, the Magpie, the Orpington and the Welsh Harlequin. The later 'heavy ducks' have often got more than a trace of Pekin in their make-up. They include the Blue Swedish, the Saxony and the Silver Appleyard. These are just the ones standardized in Great Britain and should also include the two small breeds originally named after Reginald Appleyard.

*Fig 5 William Cook's new breed: a Buff Orpington duck and drake, bred and owned by Mr Gilbert. Each first at the Dairy and the International at Crystal Palace 1908. (*The Feathered World, *1908 Vol. 39, No.1015)*

Part 1

Breeds of Ducks: History and Characteristics

1 Classic Ducks: The Old Established Breeds

The Hook Bill

The Hook Bill was mentioned in Willughby's *Ornithology* of 1678:

> . . . as very like the common duck, from which it differs chiefly in the bill, which is broad, somewhat longer than the common duck's, and bending moderately downwards, the head is also lesser and slenderer . . . it is said to be a better layer.

This description was written well before the acceptance of the Aylesbury or Rouen. William Ellis, in *The Country Housewife's Companion* (1750) wrote:

> The common white duck is preferred by some, by others the Crook-Bill [Hook Bill] Duck, some again keep the largest of all ducks, the Muscovy sort; but the gentry of late have fell into such good Opinion of the Normandy [Rouen] Sort that they are highly esteemed for their full Body and delicate Flesh.

The Hook Bill was found in the Netherlands, Germany and in the old Soviet Union, and was reported by A. Buhle (1860) as being widespread in Europe but particularly in Thuringia, where it was kept on garden ponds for its delicious meat and egg-laying capabilities.

Schmidt (1989) says that Durigen referred to the breed as *Haken-oder Bogenschnabel Ente* (Hook- or Bow-Bill Duck), *Anas domestica adunca*, where the shape of the beak is more down-turned than that of the domestic duck. The Hook Bill may have been created so that it could easily be distinguished in flight from the Mallard. Hunters could then refrain from shooting the birds on their way home at night when the wild Mallard were culled. However, to make their quick identification in flight more certain, the White-Bibbed Hook Bill was also created.

Harrison Weir wrote that the breed was said to be of Indian origin. The Dutch Historian van Gink (1932) also said that they were from the Far East and there is acceptance in Holland that the most likely place of origin is East Asia, from where they were taken and brought to the Netherlands by Dutch seafarers in the 'Golden Age'. K. Broekman (personal communication), who works in the oil industry and has travelled extensively in the Far East, has looked many times for these birds, so far unsuccessfully. One day, he asserts, he will find them if they exist.

Fig 6 Anas rostro adunco, *The Hook-bill'd Duck. (Illustration from Willughby, 1678; photocopy; by permission of the British Library)*

Hook Bills in Holland

Writing in *Waterfowl* (1987), Broekman said that Hook Bills were once kept by the hundreds of thousands in the province of North Holland (Bechstein and Frisch, 1791). The Dutch method of keeping ducks did not actually cost anything; this particular duck went off in the morning to the rivers and canals to find food and returned home before dark to spend the night and lay eggs. These domestic ducks were expected to fly quite a large distance to find their own food, so it could not be a heavy duck.

Edward Brown wrote in detail about the Dutch egg-production industry in the Landsmeer district of North Holland, in the early 1900s. Dairy farmers in the area hatched and reared the ducklings but the 'plants' where the eggs were produced were highly intensive. The ducks themselves did not show a trace of the Indian Runner type, but were rather small bodied and level from front to back:

> Some closely follow the Mallard or wild duck in coloration of plumage and markings. A considerable proportion, however, show a white cravat or crescentic marking across the throat, in which case the plumage is black or very dark. Some specimens with Mallard marking carried a very long bill, which curved downwards – the upper and lower mandibles being alike in this respect. These were reported as most prolific layers.

Brown appears to have seen the North Holland Hook Bills being used as commercial ducks in these intensive farm systems, where the birds were kept in yards bordering on the canals where they had continual access to water. The birds were fed on maize and small fish caught in the Zuider Zee, which were fed live to the ducks and were said to indispensable for the ducks' diet.

Unfortunately, in more recent times, word was spread that paratyphus (salmonella) was passed on through duck eggs. Broekman comments too that the water became polluted, and five years ago he says (in 1982) there were only thirty Hook Bills left. Many buyers chose to eat poultry meat instead of duck and, at the shows, there was a lot of competition from other 'ornamental' waterfowl for, in Holland, wildfowl such as Mandarins are still shown. This small number quoted was the Bibbed Hook Bill with the white heart of feathers on the breast rather than the dark wild colour, which has been crossed many times with other birds (Ringnalda, personal communication).

Broekman also reports that through the efforts of the Dutch Domestic Waterfowl Association the breed is back again in its full glory. At *Ornithophilia* 1987 (Utrecht) twenty-seven Krombekeenden were on show, and in 1988 the Hook Bill was allowed the 'introductory process'.

Hook Bills in Britain

Weir (1902) described birds imported from Holland, which were kept on the lake at Surrey Zoological

Fig 7 These North Holland White-breasted ducks are bred to be egg-layers rather than fliers like the Hook Bill. Their bill is also straighter. Some sources claim that the Hook Bills originate from the North Holland Bibbed and although most 'hook billed' stock was disposed of, some breeders kept them and bred them in order to create a new breed. This photograph from 1916 shows the trap-nesting system used to record how many eggs were laid by each duck and so allowed selective breeding. (Photograph courtesy of Avicultura, September 1994)

Gardens between 1837 and 1840 when 'they were the ordinary colours, mostly being white or splashed with red, yellow and brown or grey. The carriage was somewhat upright, and the necks and bodies long and narrow'. He also commented that years after, far better birds were shown at Birmingham; these were white with clear orange-yellow bills, shanks and feet, and had a top-knot towards the back of the skull. They weighed about 6lb (2.7kg). These perhaps belonged to Bakers (of Chelsea – *see* the Black East Indian) who exhibited their Tufted Hook Bills at the first West Kent Poultry Exhibition in 1853 (Hams, personal communication). Wingfield and Johnson (1853) also noted that White Hook Bills, imported from Holland, were the type most often seen.

The breed must have become rare in Britain; they do not appear to have been mentioned in later show reports. A plea to readers of *The Feathered World* in 1913 to find the 'old Dutch Hook-billed duck' drew only the suggestion of the Muscovy, depicted with a rather hooked end to its bill. The original correspondent patiently produced two sketches of the skull of the Mallard and Hook Bill to illustrate what he meant (Staveley, 1913):

> I enclose a rough sketch of the skull of (a) a 'wild duck' with a straight dished bill [sic] and high cranium, and below it (b) the sketch of 'Dutch hook-billed duck' with hooked bill and flat cranium. The latter breed were common in Holland up to 1870, and were known as 'everyday layers'. Are they quite extinct? Or could a few be found? Perhaps you have some readers in Holland who could rediscover the old utility breed.

As in Holland, the Hook Bill has had a chequered career. A white variety was kept in the UK at Kew Gardens and a few were kept by Reginald Appleyard in the Mallard colour, or in black(?) with a white bib. A few small, White Hook Bills still survived in the 1940s on a moat round the house on a farm at Metfield Hall (J. Hall). The breed was re-introduced to Britain around 1985 and again about 1993 by Tom Bartlett who brought in the Dutch standard for recognition by the British Poultry Club and BWA in 1997.

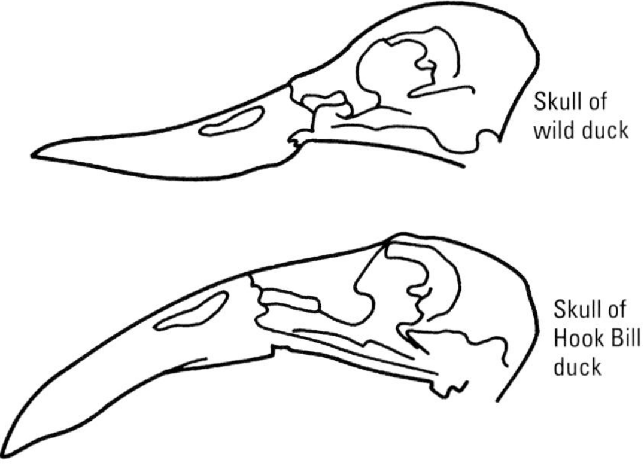

Fig 8 The skull of the Mallard compared with the Hook Bill.

Characteristics of the Breed

The Hook Bill is characterized by a slender head with a rather flat skull. In profile, the bill, head and upper neck are curved, approaching a semi-circle in shape. The neck itself is carried vertically. The body has a well-rounded, almost prominent breast yet is long and carried fairly upright at 35–40 degrees. Weights range from 3¾ to 5lb (1.8–2.3kg) in the duck and 5–5½lb (2.3–2.5kg) in the drake. Note Broekman's comment that the birds should not be too large. The modern varieties in Schmidt seem to be restricted to the Dunkel (dark or dusky) and Wildfarbig (natural, wild Mallard) colour, which may also have a white bib. In addition, there is a crested variety. The white variety has been recently re-created on the continent by using the White Campbell in the 1980s when eggs exported from Britain were used.

The British standard also restricts itself to the

Hook Bill Characteristics

Colour	
	Dark Mallard *Drake*: green-black head; absence of neck ring; body feathers mainly steel blue; speculum dull brown. Bill: slate green. *Duck*: Similar to Mallard but no eye stripes; speculum dull brown. Bill: slate grey. White Bibbed Same as Dark Mallard except for white, heart-shaped bib and white outer primaries. White White plumage; white or flesh bill; orange legs.
Shape	Slightly upright; medium in proportions; bill strongly curved downwards.
Size	Light: 3¾–5½lb (1.8–2.5kg)
Purpose	Utility egg-layer

Fig 9 Head of a Hook Bill duck exhibited by Richard Sadler, Stafford, in 1998. She shows the almost semi-circular shape of the head and neck in profile. The breast has a neat, white bib.

Dark Mallard, the White, and the Bibbed varieties. The Dark Hook Bill shows 'dusky' characteristics, in that the duck lacks the typical Mallard eye-stripes and the bright blue speculum of the secondary feathers. The better continental birds have a rich ground colour and clear pencilling on the larger feathers. In the Bibbed varieties the margin of the bib should be clearly defined. The bib itself is white and is accompanied by two to six white primary feathers.

The Aylesbury

Everybody has heard of the Aylesbury duck. Its fame has spread all over the world. Point to a large, white duck in the farmyard and some one will say authoritatively: 'That's an Aylesbury!' But the chances are that it is not. Most farmyard white ducks have their origins closer to Beijing than the English town in the Thame valley. If the bird has an orange bill, a smooth breast and stands more than slightly upright, it has Pekin blood in it and is nothing like a true Aylesbury.

The pink-billed Aylesbury was the mainstay of the English duck industry until well into the nineteenth century, when it became the victim of its own success and eventually was little more than a trade name, like *Thermos* and *Hoover*. *Aylesbury* became the synonym for any big, white duck. Then it was ousted in popularity by the influx of new foreign breeds, namely the Indian Runner, which had a phenomenal track record for egg-laying, and the large, cream-coloured Pekin, which could be crossed with the Aylesbury or the Rouen to produce more fertile and vigorous offspring. By the final quarter of the nineteenth century, the real Aylesbury had almost disappeared, whilst its name lived on like a ghost to haunt its virtual extinction.

The Aylesbury Duck Industry

For the interested reader, Alison Ambrose' book *The Aylesbury Duck*, published by the Buckinghamshire County Museum, is a fascinating account of the social and economic effects of the duck industry in that area. Even in the 1750s, 'the poor people of the town are supported by breeding young ducks; four carts go with them every Saturday to London'. It was the buying power of the rich in the capital that provided the market for the industry. The impetus for further growth came with the railway. The railway companies or agents even collected the ducklings (aged six to eight weeks) from the village homes for transport in 'flats'. These were wicker hampers in which the dressed carcasses of the birds were packed.

Small duckers reared perhaps 400–1,000 ducks a year but large breeders, like the Westons and the appropriately named Fowlers, sold many thousands. The small duckers were the labourers who had saved enough capital to start rearing the ducks, for they generally bought the eggs from the breeders. The stock ducks were kept on the farms where the conditions (for the ducks) were healthier, to promote fertility and hatchability. Many of the eggs were hatched by the cottagers who lived in the town and who had town jobs.

One part of Aylesbury was known as 'Duck End' and accounts of it at that time indicate it to have been a slum. It suffered from the cholera epidemic in 1832. This was not surprising because of the sewerage of the town terminating there and the existence of a large number of ducks and other animals, which were kept in the houses and yards of the area. It was so wet that the sewage from the duck ponds in the back yard of one house passed under the floor of the living room, causing the soil to ooze up through the bricks. Ambrose quotes the Rev. John Priest from 1810 to give the flavour of a ducker's cottage:

> In one room belonging to this man (the only room he has to live in) were ducks of three growths, on 14th of January 1808, fattening for the London market: at one corner about seventeen or eighteen four weeks old; at another corner a brood a fortnight old; and at a third corner a brood a week old. In the

bedroom were hens brooding ducks in boxes, to be brought off at different periods.

As you can see from this account, the table ducks were not permitted swimming water and were not even permitted out of the 'purlieus' of the dwelling house till they were killed. They were fed on chopped, hard-boiled eggs, mixed with boiled rice. This was given to them several times a day and 'their clamour as feeding time approaches is terrible'. Bigger producers fortunately kept them in sheds covered with barley straw and the birds were fed barley meal and bullock's liver as well, though the less reputable producers were accused of using horse flesh and even carrion. Fowler's account (in Wright) pointed out that this intensive rearing was only the strategy employed for producing the table ducks, which were generally killed at about six to eight weeks old and weighed about three to four pounds each.

Yet the days of the specific Aylesbury trade were numbered. After 1890, the increase in demand for ducks was not met by the original producing area, but by large-scale industries, which developed in Lancashire, Lincolnshire and Norfolk. In addition, there was the challenge of the Pekin. These ducks had been imported from China in 1873 and, having no bagginess of flesh on the chest, soon gained favour as a commercial cross.

Exhibition Birds

Whilst the Aylesbury Duck industry was at its peak in the second half of the nineteenth century, the Victorian exhibitions also gave the breed additional prestige. At the first National Poultry Show for live specimens held in the Zoological Gardens in 1845, there was a class for the 'Aylesbury or any other white variety' and, soon after that, the major agricultural shows began to include a poultry section in their schedule of events.

Birds intended for exhibition, or egg production on the farms, were reared using methods still more akin to the exhibition breeders today, for care was needed to produce the best. Ordinary ducks weighed only 6–7lb (2.7–3kg) at a year old, whereas the heaviest pair exhibited by Fowler at Birmingham weighed 20lb (9kg) in the late 1800s. Weighing by the pair was customary at the shows in those days and this led to elaborate preparation. Ducks were fattened to the state of obesity; there are records of birds being stuffed with worms and sausage meat, sometimes with dire consequences, such as death from 'apoplexy' in the show-pen.

In early illustrations of the Aylesbury, such as that

Fig 10 Aylesbury Ducks bred by Mrs Mary Seamons of Hartwell: cup at Aylesbury 1870 and many other cups and prizes. (Outline drawn from Ludlow's painting in Lewis Wright, 1874)

of the artist Ludlow in Lewis Wright, Mrs Seamons' prize-winning pair of ducks did not show a keel. This seems to have been developed as an exhibition feature, so that by the late-nineteenth century there came to be two separate strains of the duck – one for the show-pen and one for the table. This did not always meet with approval. Harrison Weir (1902) wrote that the modern Aylesbury of 1902 'has been so much increased, though not in beauty, that it now has the appearance of a body inside a feather-covered skin that is at least a size too big'.

Despite this adverse comment on the exhibition Aylesbury, Weir did concede that the Scottish breeder, Gillies, grew drakes at six months old, which reached 12lb (5.4kg). It was this type of bird that was now to dominate the show-pen. In the 1920s and 1930s, Huntly and Son (with their Coldstream type), together with the Westons, won at the Palace Show. New names arrived. Vernon Jackson improved his own Aylesbury strain with the MacBean line from Cornwall in the 1930s and, together with Nick Thomas, these two breeders continued the exhibition line post-war as well. At 85 years old in 2000, Mr Jackson can look back on nearly seventy years with this breed.

Exhibition Aylesburys are huge, oblong-shaped ducks in white plumage. The 1901 standards specify that six-month-old ducklings should not be less than 9lb (4kg) for the duck and 10lb (4.5kg) for the drake. 'The second year and afterwards, the duck should

Fig 11 White Aylesbury Ducks. The property of Captain Hornby of Knowsley Cottage, 1852. These were selected as typical Aylesburys for The Poultry Book *(Wingfield and Johnson, 1853). They had pink bills, pink shanks and feet. (By Harrison Weir)*

*Fig 12 The Aylesbury in 1905 for comparison. The Countess of Home's Aylesbury drake. Winner of the first and challenge cup International (Alexandra Palace) and first and challenge trophy Crystal Palace in 1905. (This was the Coldstream strain.) (*The Feathered World, *March 1907)*

equal the drake in weight, and neither should be under 11lb (5kg). Anything over these weights to count extra merit.'

Such a massive body needs strong legs and feet set in the middle to support the weight, but these birds are surprisingly agile for their rather unwieldy appearance – as long as they have exercise. The body carriage is horizontal, with the shoulders slightly raised when alert in a show-pen – a characteristic liked by Vernon Jackson in fit ducks. Aylesburys should not droop down or sag at the shoulders; the bird should be able to manage the keel, which starts on the breast and runs under the body, touching the ground. One should get the impression of an oblong duck that is neither cut away, nor baggy, at the back. The body is so big that the wings have to fold neatly at the sides rather than cross over at the rump.

The head is a very important feature of the exhibition bird. It should measure 6–8in (15–20cm) in length from the back of the crown to the tip of the bill. The top line of the bill is dead-straight and it should join the head smoothly without an abrupt rise to the crown. An old, apt description was that it should look like a woodcock in profile.

Despite being a white duck, colour is important in breeding too. The legs are bright orange, though in Weir's colour print (Wingfield and Johnson, 1853) the legs are the same colour as the bill. The bill must be 'as pink as a lady's fingernail', a requirement once thought only possible to meet in the Thame Valley, where the river gravel wore off the top layer of skin as the birds dabbled in the river. However, it is now realized that putting gravel in their water troughs, and keeping them away from too much grass and sun, will maintain the desired exhibition colour.

The exhibition Aylesbury is not easy to breed, but some of the reasons for Vernon Jackson's years of success can be identified. Young, fit drakes without too much keel (which gets in the way for mating) should be used with mature (two- or three-year-old) ducks in the ratio of 2:5. The birds, according to Mrs

Fig 13 Young Aylesburys at Tom Bartlett's in 1999. The keel on the breast is well-developed and the underline of the bird is parallel to the ground. The birds have a long, pink, straight bill – essential in this breed. A black bean sometimes develops in the females with age, but should be avoided in the breeding stock if possible.

Seamons too, 'are very fond of green food and . . . plenty of room'. Perhaps most telling is Edna Jackson's quip 'Half the pedigree is by mouth'. As with all heavy ducks, size and quality is only achieved with attention to diet at all stages of production.

Aylesbury Characteristics

Colour	White plumage; pink bill; orange legs and webs
Shape	Long, horizontal and deep body, keel parallel to the ground; long bill
Size	Heavy: 9–12lb (4.1–5.4kg)
Purpose	Table bird originally; exhibition

Fig 14 Head of an exhibition Aylesbury drake. Paul Meatyard's winning Aylesbury at Stafford, 1999. The eye is fairly close to the top of the head, but not as high in the skull as in a Runner.

The Rouen

Origins

Rouen, Rhone or Roan: these have all been offered as the breed's original name. The latter two could easily be the result of mispronunciation or corruption of the more acceptable 'Rouen', yet each has a logic of its own. Roan is used to indicate the intermingling of grey and colour so prevalent in the breed. Rhone is, of course, another area of France

Fig 15 Richard Waller's traditional commercial Aylesbury Ducks were recently featured in Poultry World. *His birds and the business belong to a family tradition that goes back over 200 years and the ducks still show the huge Aylesbury head and the pink beak, but the rounder and tighter breast of the table strain. His is probably the last of the original Aylesbury businesses. Birds such as these could be used to re-invigorate the exhibition line, which is becoming increasingly difficult to breed.*

famous for its cuisine and wines and the term 'Rhone' does seem to have been used as a name for the duck in 1815. However, many authors acknowledge that these large ducks were successfully selected in Normandy, hence the name of the regional centre, the city of Rouen. Even in 1750, William Ellis reported that 'the Normandy Sort . . . are highly esteemed for their full Body and delicate Flesh; they are great Devourers of Grain . . .'. In *A Treatise on Poultry*, Fuller (1810, quoted in Weir) wrote of the Rouen:

> The large, fine species that answers so well in the environs of Rouen on the banks of the Seine, on account of its being in the power of its keepers to feed them with earth worms taken in the meadows, and which are portioned out to them three times a day under the roofs where they are cooped up separately.

Weir stated that this indicated that the Rouen duck was 'merely the wild duck enlarged by domestication and high feeding'. Many of the best Rouens in the south of England had a French origin. He quotes (no reference) that in the period 1800–10:

> . . . ducks are therefore a trade of the coasting captains of this nation (the English) who, in passing to return home, sell again to rich landowners wise enough to reside in their domains. The profit of exporters depends on fair weather and shortness of passage which keeps off, more or less, a mortality among their passengers [ducks].

The author was therefore dubious that the import of Rouen ducks in the 1840s by Bakers of Chelsea was the novelty they made it out to be. One of Harrison Weir's pictures of the breed (Wingfield and Johnson, 1853) shows the birds to be little different from the Mallard in colour but about three times larger than the Black East Indians next to them in the print, placing them at an estimated 6lb (2.7kg). In the same volume, the authors refer to drakes reaching 6lb each at only ten weeks old.

The breed was developed to exhibition size and status in Britain. Ludlow's painting of a Rouen pair in Lewis Wright's Victorian *Book of Poultry* shows specimens with less well-developed keels than large birds possess today. The duck is rather dark by today's exhibition standards and although some double chevron marking can be seen on the flank feathers, the wing coverts remain similar to the Mallard's.

Characteristics of the Breed

Plumage of the Duck
When looking for a really good bird, there are five areas on which to concentrate: shoulders, wing coverts, rump, chest and head.

Shoulder (or scapular) feathers demonstrate the essential characteristics. The big feathers should have a rich golden brown colour. Depending on the strain, and also the effect of the sun, this ground colour can range from what the standards book calls 'almond' to deep chestnut. Within the 'ground' should be two clearly defined chevrons of almost black lines. The smaller feathers tend to have a central line surrounded by a single chevron. Above all, these should be clear. The poorer birds will be dark, drab and indistinctly pencilled.

The smaller wing coverts too should be clearly marked, each tight feather being clearly laced to give a 'honeycomb' effect. The greater coverts, which overlie the secondary feathers (which carry the blue speculum), should also be clearly marked with distinct bands of black and white.

On the rump, the poorer birds will have simply a dark or fuzzy 'mass' of plumage. Good birds will have contrast and clarity in these minor feathers. In the show-pen these qualities stand out blatantly. Look then at the chest. Clear un-blurred markings are the things to go for.

The head is the final key area for colour. Sometimes, Aylesbury blood has crept into the strain. It may have been useful at some time to reinvigorate the breed, add size or perhaps give better ground colour, yet it has had numerous negative effects. It is seen most clearly in the bill colour of the male but also in the female head. She should have an orange bill with a dark saddle extending over the sides and roughly two-thirds towards the tip. Aylesbury blood can add a pink paleness and a lack of saddle. Over the course of a year the bill colour will tend to change somewhat, going quite dark in the summer months, yet a constant, overall slate colour should also be avoided. The head plumage resembles a female Mallard. There should be a dark line that seems to go 'through' the eye itself, as well as the broad, dark band that extends from the neck to over the crown.

Plumage of the Drake
The drake's plumage is beautiful and strikingly similar to that of the Wild Mallard. The rich claret bib on the drake should be a solid and clearly defined patch of colour. Any fine, pale fringe to the feathers is a fault. Young drakes often show this

Fig 16 Good definition of the lacing on the smaller wing coverts and a well-defined white bar are prized in a Rouen duck. She should also have clearly defined dark pencilling on the larger feathers, seen in the foreground.

Fig 17 Head of a typical Rouen duck showing the Wild Mallard face markings – pale lines above and below the eye, and a darker stripe through the eye. The bill also has the darker central saddle and dark bean.

defect when they first grow their adult plumage, but the claret colour often intensifies and the fault clears later in the autumn. The bib's outline should be clear and sharp, neither blurring into the grey flank feathers nor trailing down towards the leg coverts as in the Silver Appleyard.

The standard Rouen drake's grey feathers should persist all along the flank till they join the black rump. Quite often there are several light-edged feathers in the grey flank feathers close to the stern, but this is regarded as a fault.

Spinke (1928) used the term 'chain armour' or French grey where each of the drake's grey feathers is 'pencilled' with wavy lines. This pencilling is quite different from the pencilling on the duck, where the dark lines are concentric in form, following the outline of the feather. In the drake, these dark lines cross the feather (another definition of pencilling). The lines appear wavy and even discontinuous, as if stippled with a stiff paintbrush; they are the standard body feather pattern of the male Mallard.

The back, rump and under-tail cushion are a rich green-black, whilst the tail and primary feathers are a dark brown slate. The secondaries carry the speculum as in the duck.

Size and Shape

Apart from colour, the Rouen also has to have the right size and shape. Wright described the birds as 'massive'. They must be a big duck in the show-pen reaching 9–11lb (4–5kg) in the female and 10–12lb (4.5–5.4kg) in the drake. The long, wide bill must be in proportion to the strong head. The body is long and broad with a deep keel. They are oblong in appearance but when moving in the field have tremendous strength and get about very easily.

Breeding for Colour

In the 1920s, there was a trend for double-mating in the Rouen, described by Spinke in *The Feathered World* (1928). This meant that two breeding pens were kept. One was for breeding richly coloured males with a strong claret breast and no white-edged feathers around the stern. The female counterpart was rather dull but, for the 'drake pen', these females were ideal.

The emphasis for exhibition on the chestnut females with sharp, even pencilling, meant that lighter coloured males were used for breeding. It is when drakes are in their juvenile feathers that they are selected as breeders for next year's 'duck pen'. This is because a male in his 'duck feathers' at ten weeks old shows the pencilling and ground colour of the brightly coloured females that he has the potential to produce.

Fig 18 Feather markings in the Rouen. The larger feathers of the duck show double-pencilling in dark brown on the lighter ground (left). The drake's feathers also show 'pencilling' of a different type. Dark lines go across the feather giving a stippled effect (right).

Fig 19 Breeding-pen of two Rouen drakes and four ducks.

Exhibition

To produce any pure strain, a certain amount of inbreeding is necessary but this continued practice, which restricts the gene pool, eventually restricts viability. In the 1930s pure exhibition Rouens were becoming very difficult to breed, but they were a favourite at the exhibitions. Show results from the Crystal Palace and the Dairy Show in 1930 included Spinke and Alty, who specialized in this breed for many years.

By the 1950s, Rouens were nearly impossible to reproduce (Jackson, personal communication). Appleyard too commented that the exhibition birds were not very prolific; with continuous inbreeding, the eggs are simply infertile. Import initiatives in the 1970s revived the fortunes of the breed. So, although some of the older breeders do not think that Rouens are as beautiful as they were before 1940, at least they do reproduce. John Hall's birds dominated the show-pens in the 1980s, a typical duck being illustrated in Roberts' *Domestic Duck and Geese in Colour* (1986). Such birds provided the foundation stock for many breeders during that time.

Breeding

Rouens do not usually lay as early as the Aylesbury, which, in a young duck, starts in February. Rouens usually bide their time until March and, quite frequently, these early eggs are infertile. This does not matter to the pure-breed enthusiast. The best Rouens are often hatched from the later eggs laid in late April or May. The disadvantage of hatching late, of course, is that the birds take a long time to mature, so that it is difficult to select the best for retention. Rouen drakes take over twenty weeks to gain complete adult feathers and, after that, they can take a year to complete their size. This is especially true of the exhibition strains.

Because of their colour and slower gain in weight, Rouens have never figured in the duckling trade but were traditionally used for the autumn market. The early slaughter age of commercial ducklings now means that Rouens are totally uneconomic. They are mainly kept as very ornamental, useful birds in Britain but nevertheless stock should be selected with care. Size and colour should be preserved, and also shape. The standard says the body should be long, broad and square. Certainly birds that look oblong in profile seem healthier than those with over-rounded or 'roach' backs. Older birds tend to droop at the shoulders and at the rear but young ones must not have these faults. The birds should be kept fit. Plenty of grass, exercise and a stream will ensure that they breed.

The Rouen Clair

Although this breed may be as old as the Rouen, it has rarely been described, perhaps because it has not been exhibited a great deal in the show-pen. The first official standard in Britain was in 1982, when a number of breeds of ducks and geese, many of which had been in collections for over a century, were finally admitted to the Poultry Club Standard.

This revival of interest in standards was after a period of great difficulties for the exhibitions in the early 1970s. The International Poultry Show and the Dairy Show at Olympia had both closed down, and the devastating fowl pest outbreak of 1971 severely restricted poultry movement. However, following a visit to the Hannover Show by Poultry Club members, the Club set up its own National Show at the Royal Show Ground at Stoneleigh. There was an enormous influx of members – both old and new – which resulted in the Poultry Club administration being kept busy 'approving, adapting and revising standards to permit these old and new breeds to be classified for exhibition. Happily, waterfowl too figured in this resurgence. . .'(Will Burdett, 1979).

This resurgence resulted in the British Waterfowl Association producing standards for the newly imported Trout and Mallard Runners, plus the Blue Swedish, Harlequin, Appleyard, Saxony and Rouen Clair. There was also a revival of the Blue and Black Runner standard and the Decoy (Call), a breed that was about to make a huge come back. Burdett further commented '. . . what a wonderful addition a full colour WATERFOWL STANDARDS would make to the list of BWA books!'.

The Rouen Clair Standard

Nobody really knows the true story of how the Rouen Clair was produced in France. Somewhere the Rouen had been crossed with birds that introduced new colours. This could have happened before 1853 when it was spotted that there were two colours of ducks imported from France,

> the one with plumage like that of the Rouen, the other a much darker bird, both the duck and the drake being of a very drark brown, almost black, each having a white mark on the front of the neck, but not encircling it

The latter could have been an early sort of Duclair, a white-bibbed, dark bird. There was probably no specific reference to the Rouen Clair until Charles Voitellier (1909) recognized not only the Rouen and the Duclair, but also the Canard de Rouen, *variété clair*.

It seems unlikely that our modern birds go back directly to this stock. van Gink, writing in *The Feathered World* in 1933, referred to the 'new French Rouen duck' from Picardy:

> In France, the Rouen ducks are bred in two colour types nowadays: dark (*foncé*) and light (*clair*). The light-coloured Rouen is a new production, bred in order to get a purely utility duck with all the beauty and size of the English Rouen duck
>
> It was a certain Monsieur Rene Garry who produced this new type of Rouen duck by crossing and careful selection . . . [he] selected the breeding stock, to start with, from ordinary Rouen ducks, as they were found on the farms of Picardy, which are the lighter colour type, the ground colour being of a pale buff, a colour usually called 'isabel' on the Continent . . .
>
> We have carefully studied these light-coloured Rouens at the Paris shows, and liked them very much, but could not get away from the impression that Pekin blood had been introduced into France at some time. The shape of the head is too Pekiny not to be related to this wonderful Asiatic breed and we all know that this type of head was unknown in Europe before the Pekin ducks arrived.
>
> In crossing Pekin ducks into Rouens one gets brownish-black crossbreds which, to a great extent, show the markings of the Duclair ducks, so much akin to the Rouen. We are giving sketches of an almost ideal drake, seen in different ways, showing the deep and broad shape of this variety, and also sketch a winning duck drawn from a photograph, which plainly shows the thick, rather short Pekiny head.

The 1982 standard gave only slightly different weights from Brown's description, and stressed that the breed 'must remind one of the Common Mallard – a body long and well developed'. Like the Wild Mallard female, the duck must have the characteristic Mallard eye stripes, and the paler throat. The drake, too, looks like a large Mallard. He has a white collar, which fails to encircle the neck completely, a mulberry bib, typical Mallard-pattern body feathers and glossy green-black rump. Unlike the Rouen, the stern must be white. Both sexes have a brilliant speculum – indigo rather than blue – and the duck is more highly prized if she has well-defined pencilling in the form of a chevron on the larger feathers. Today, the beak colour is similar to the Mallard, an orange-ochre bill with a dark saddle being preferred to grey in the female. The bill is yellow with a greenish tint in the male.

Fig 20 Three outline sketches of the Light Coloured French Rouen Drake. (van Gink, 1931)

Fig 21 Head of a Rouen Clair female showing the typical, but exaggerated, face markings of the Mallard, i.e. the darker stripe of feathers through the eye, and the paler feathers above and below. The throat is paler but these paler feathers do not continue onto the breast. (Heather Birkett, Stafford Show, 1999)

Fig 23 Rouen Clair drake showing typical markings. His white collar almost encircles the neck and the stern of the bird is white. These birds should be large, weighing up to 9lb (4kg). (M. Hicks at the Poultry Club National, 1999)

Fig 22 A group of Rouen Clair females. (Tom Bartlett's, August 1999)

Utility Ducks

These birds have never really been developed for the show-pen in large numbers. The Rouen Clair has probably always been very variable because it was designed as a utility farmyard type. Because the duck can be made from crosses, this explains its variability and also its rapid growth as a table bird, plus its capacity to lay a lot of eggs. Cross-breeding produces big, healthy ducks, which can be prolific. The breed is perhaps an early 'designer duck' – a deliberate or accidental cross that resulted in a useful utility duck. Dave Holderread (1985) notes its variability too, because the duck has long been bred for its utility qualities, not its appearance. His ducks typically lay 170–200 eggs and grow to 7–8lb (3–3.6kg); the largest drakes reach 10lb (4.5kg).

Call Ducks

Call Ducks have instant attraction. They have appealing baby features to which we are programmed to respond: round faces, big eyes and diminutive size. Apart from being visually attrac-

Rouen Characteristics

	Rouen	Rouen Clair
Colour	Developed from the wild Mallard	A lighter version of the wild Mallard
Shape	Horizontal; keel parallel to the ground; long and deep body; long bill	Slightly upright (10–20 degrees); long body; little evidence of keel; medium bill
Size	Heavy: 9–12lb (4.1–5.4kg)	Heavy: 6½–9lb (2.9–4.1kg)
Purpose	Table bird originally; exhibition	Table bird

Fig 24 Two exhibition-quality white Call Ducks showing the round head, bright eye and short, broad body.

tive, Calls can also become very tame. They have an in-built blasé attitude towards people. Buy a pair of young Calls, perhaps reared with a large number of other birds that have never received much attention, and they will be rather wild initially. But keep them in your back garden for a few weeks and they will end up quacking at the kitchen door.

Standard Features of the Exhibition Call

Although people often say to breeders that they do not want expensive show birds, they only want pets, it is the standard features of the exhibition Call, which are difficult to breed, that make it so attractive. The birds have a broad, deep body and when viewed from above (on judging them) the body is short. Any birds that have a long or thin, boat-shaped body are rejected. The legs are set about halfway along the length of the bird so that they usually have a horizontal carriage.

With the round body goes a round head with a high crown, set on a short neck. The large, bright round eye is set nearly central to the face, a bit above the half-way line. This gives the bird its baby-like look. The beak should also preserve the same quality. All newly hatched ducklings have instant appeal because, proportionately, their beaks are smaller in length than the adults, even in Runners. As the ducklings grow, the beak lengthens. In baby Calls the beak is minute, often as broad as it is long in a good show strain. As the birds grow, some of these beaks will lengthen but hopefully some ducklings will become the show birds. The standard bill length is a maximum of 1¼in (3cm) but good exhibition Calls have even shorter bills. More important than the actual length, though, is the proportion of length to width, which should be 2:1. A wide bill appears to be shorter than it actually is. A bill that is set at 90 degrees to the skull will also appear to be longer. Lou Horton (1995) suggests that the symmetry of the head, and the bird as a whole, is better when the bill is set 5–10 degrees lower.

The size of the birds is important too. In Germany, the stated weight of the *zwergente* (dwarf duck) is no more than 1kg (2.2lb) (Sheraw, 1983). Standard weights in Britain are 1¼–1½lb (0.6–0.7kg) for drakes and 1–1¼lb (0.5–0.6kg) for ducks. Birds that greatly exceed these weights should be penalized, but it should be borne in mind that birds kept on different regimes vary immensely in weight. Free-range Calls kept in a fox-proof pen and on ad lib food are solid and exceed the standard weight, whereas poor, thin birds will be at the standard weight even if they have a large body frame. Judges always have to use their discretion in interpreting the standard. Tiny birds that are underweight should not be favoured. They often lack the fullness of body required and are sometimes useless for breeding.

Where Did They Come From?

Calls are often referred to as Dutch Call Ducks and even as the Dutch Dwarf Duck; they are certainly the most popular domestic waterfowl breed in Holland. Land is scarce and expensive so few Dutch keep geese, but the popular Call is small and easy to keep. Nobody disputes that both American and British exhibition stock had Continental origins. Dutch Calls were in England before 1851 and must have been in the USA before 1884 as Whites and Greys (Mallard) have a detailed description in the first American Poultry Association *Standard of Excellence*.

However, Dutch breeders feel that the Dutch

Fig 25 Head of a baby Call Duck. The ducklings are very small and fragile, as can be seen from the finger tips, for scale.

Decoy, referred to as the Call by the English, was originally Asian. This is not surprising for two reasons. First, Holland was a colonial power; Indonesia was formerly the Dutch East Indies and the Indian Runner certainly came to both Holland and Britain from this source. Second, the ancient civilizations of the Far East had had thousands of years to develop their local strain of the Mallard – as evidenced by the record of the Runner – and it would be no surprise if the Eastern people, experts in aviculture, had produced the Call as well as the Runner and Pekin.

It is not known for certain how long the Decoy Duck has been bred in Holland. Writing in 1678, Willughby described how Coy-ducks were used to catch wildfowl.

> The Coy-ducks are to be fed at the mouth of the entrance of the Pipes, and to be accustomed at a token given them by a whistle to hasten to the Fowler. The Fowler first walks around the Pool, and observes into what Pipe the birds gathered together in the Pool may most conveniently be enticed and driven, and then casting Hemp-feed, or some such like thing at the entrance thereof, calls his Coy-ducks together by a whistle. The wildfowl accompany them, and when the Fowler perceives them now entered the Pipe he shews himself behind them through the interstices of the hedges . . ., which being frightened, and not daring to return back upon the man, swim on further into the pipe. . . . The Coy-ducks go not into the cylindrical net, but stay without and entice others. . . . Of the Coy-ducks some fly forth and bring home with them wild ones to the pool, others have the outermost joint or pinion of their wings cut off, so they cannot fly, but abide always in the pool.

The idea of the Decoy (*de Kooi*) Duck and the pipe or decoy trap was invented by the Dutch, but it is uncertain that these early Coy Ducks were themselves the Call. Although sufficient time had elapsed by 1678 for Dutch traders to have brought ducks back from the Far East, these Decoys may have been Coy Ducks by training rather than breeding.

Schmidt notes that Calls seem to have become popular in Holland round about the year 1800. They became widespread within quite a short period of time. This indicated that they were not developed in Europe, but arrived 'ready made' from the Far East. It was surmised that they had probably been kept for centuries as ornamental birds in parks and courts. The source of this information may have been van Gink (1932), who suggested that since the Dutch had Japanese bantams as early as the seventeenth century, it was also possible they brought back bantam ducks as well.

> . . . we should not be surprised if some day Japanese poultry and duck fanciers might find in their old books information relating to some old breed of dwarf ducks, especially as the Call Duck's type is very different to the ordinary European type of duck to sport from it, and since they bred so true it must be a very old-established breed.

Writing in the *British Waterfowl Association* year book (1976), Cees Stapel (a Dutch breeder working in England) felt that although it is believed that the Call came from Holland, there is no direct proof of this. Calls were unknown in Holland before the Dutch East Indies were acquired as a colony. They are not shown in the paintings of old Dutch masters

Fig 26 Laysan Teal have been labelled 'degenerate and inbred Mallards'. They come from the small Pacific island of Laysan, in the island chain north west of Hawaii. The duck population was reduced as a consequence of the introduction of rabbits in 1903. It is speculated that the Japanese did take the Laysans back to Japan, to develop into the ornamental Call Duck.

up to the eighteenth century, even though the Crested Duck appears in a painting from 1660. He states that 'The systematic breeding of these [Call] ducks is much older in China and Japan than Europe'.

A further tenuous and unsubstantiated link with the Far East is the suggestion that the Laysan Teal (*Anas laysanensis*) may have been used to produce this breed. Both the Laysan and the North American Black Duck are regarded as ancient offshoots of the Mallard and 'some authorities unglamorously describe the island form [the Laysan] as insular, degenerate and inbred Mallards' (Todd, 1979). Todd notes the teal 'were subjected to severe hunting pressure by commercial guano diggers, and there is some evidence that Japanese feather hunters worked the island'. The island of Laysan itself is part of the Hawaiian island chain in the centre of the Pacific, east of Japan. It is therefore rather more accessible to the Japanese culture than the European. The Laysan shares some characteristics with the Call, such as the shrill quacking (quite unlike other breeds of domestic duck), active character and short, compact body. In addition the incubation period is the same as the Call (26 days) and Laysans are also reputed to be very tame.

The Decoy Duck in Victorian Britain

> . . . is a small variety of domestic duck, much addicted to calling; on seeing any bird on flight, they instantly call out, so as to induce its return. They are the colour of the wild duck; they take their place amongst them, and are trained to lead them into nets. And, when there, to desert them.

Nolan wrote this succinct description of the Call Duck in 1850 when these little birds were much prized by wild-fowlers at a time when the capture of the Mallard was still a commercial enterprise. Not only were the wild ducks caught for sport but they were also slaughtered, with the help of the Decoy Duck and the decoy cage, *en masse*. The word 'decoy', seems to have come from the Dutch word for 'trap' (*de Kooi*). The little bird was aptly named.

Wild duck were caught in huge numbers on lakes, particularly where they were surrounded by woodland, in large, specially designed structures. These were basically ditches about twenty yards long and four yards across at the entrance (though some were much larger), which tapered to a deadly end, only two feet wide. The edge of the ditch had to be quite steep so that the birds did not investigate the banks to rest. Instead, 'loafing sites' were provided, such as half-submerged logs. Poles were driven into the ground either side of the ditch and arched over to form a tunnel. Over the whole structure, reinforced by horizontal poles, was slung a net to make a 'pipe', which over the last four yards, screened by reeds, became a place of no escape.

The job of the Decoy or Coy Ducks, which were either half-domesticated or even tame ducks, was to return to the lake each night to be fed by their keeper. They also brought down the Wild Mallard to feed, rest and feel secure at night. So at daybreak, when the dog appeared to disturb the ducks, the Mallard could be made to swim into the decoy.

According to Kear (1993), the layout of the trap, the screens of reeds around it, and the function of the dog, were quite complex. The best dogs were well-schooled and preferably red in colour, like a fox. The Dutch even had a specific breed, the *kooikerhondje*, to act as the 'piper' (from the pipes of the decoy, not

Fig 27 The Decoy, from Nolan 1850. White Ducks and perhaps wild geese are amongst the Mallard caught in the decoy shown in the background.

the 'pied piper').The dog did not panic the birds but evoked a mobbing response. As the dog moved along the pipe, so the wild ducks followed it and swam up the tunnel towards the deadly purse end. The coyman only appeared from behind screens of reeds when the ducks were well inside the pipe. There were often several of these decoys around the same lake and they were chosen according to the wind direction to maximize the catch. Nolan relates how the Tillingham decoy in Essex netted, in 1795, 'ten thousand head of widgeon, teal and wild duck Mallard'.

For wild-fowlers who hunted alone, the Call Duck was no less useful. The hunters were said to set a tame duck on the water and attach a piece of string to her leg, so that a tug on the line evoked a loud quack when the situation demanded it. The idea was that the tame birds called the Wild Mallard down. The tiny ducks were popular because they were light and easy for the wild-fowlers to carry around the marshes.

Most of the English Decoys differed very little from the ordinary Mallard in shape and colour and were often Decoys largely by training; some did look different and were splashed with white, or were fawn. This similarity in shape to the Mallard was not the case with the imported Dutch Decoy Ducks, which were smaller and a different shape, the variety becoming known as the Dutch Call Duck.

> The head is of much rounder form than that of either the Wild or Domesticated Duck, bearing, in fact, a close resemblance to that of the Tumbler Pigeon. The bill is also wider at the extremity than any other of the Domesticated Ducks (Wingfield and Johnson, 1853).

Tegetmeir in *The Poultry Book* (1867) also described Calls as differing from ordinary breeds in that they were small and possessed a round forehead and broad, short bill. He described the Grey Call as the exact counterpart of the Rouen and wild breeds not only in plumage but in legs, feet and bill. The white 'should be clothed in feathers of pure unsullied white; the bill, however, is not flesh coloured as that of the Aylesbury, but a bright, clear unspotted yellow, any other colour being regarded as disqualifying the birds from success in severe competition'.

Calls in Britain and America may have stayed in these basic colours for decades, but we do get a suggestion from Weir that, when bred in numbers, the Dutch Calls were already producing 'blues and buffs', as he called them, as early as 1902. Weir's Dutch Calls were:

> . . . charming in both form and size, and the white, when really good, with full Dutch-boat shaped bodies, short necks, and rounded heads with short bills of deep bright orange tint, and standing on short equally brightly coloured shanks and feet have a truly lovely appearance. . . . The voice of these Dutch coys or Call Ducks is peculiar, piping and shrill . . . almost musical where a number of ducks are indulging in their evening loquacious prattle.

In contrast to Weir's allusion to colour, the American *Standard of Perfection* allowed only the White and the Grey in 1905 and Appleyard in 1949 did likewise. It seems that in Britain and America we had to wait until the 1960s for much variety of colour to be generally recognized.

Disappearance

The Call seems to have more or less disappeared from readily available literature in the years between 1902 and 1970. They were very rarely advertised for sale, appearing a few times in *The Feathered World* in the first decade of the twentieth century, and then disappearing until the 1940s. It may have been that the Call was not very popular because people found very little use for the birds. They were too small to

Fig 28 Grey Calls. The duck, in particular, shows the high crown and also the compact body of the ideal Call. (From Tegetmeir's The Poultry Book, *1867)*

Fig 29 Call Ducks. (Illustrated and owned by Harrison Weir, 1902)

eat; people were not on the whole accustomed to keeping ducks as pets; and by the early twentieth century the day of the working Decoy Duck was over. The marshlands of the Fens were drained for arable agriculture, so more grain was available for intensive rearing industries.

In Britain and America, the Wild Mallard was replaced for eating by the white table duck. The early decades of the twentieth century were also the years of the 'designer ducks'. Varieties were made for commercial purposes by using the Indian Runner crossed with the Aylesbury and Rouen. In competition with the egg-laying potential of these new rivals, the Call Ducks could not compete for commercial interest. They probably survived in wildfowl collections where they have always been kept to keep the wildfowl at home and warn of predators.

Calls even disappeared from the Poultry Standards book. They were there in the first edition of 1865 but Lewis Wright commented in 1874:

> At one time they were the principal 'fancy ducks' shown; but of late the Mandarin, Carolina, and other more striking varieties have pretty much superseded them as exhibition birds, though they still retain their popularity for lakes and ornamental waters, and are occasionally used as decoys. . . . For show they should be small as possible, but very good ones are rather rare. Mr Serjeantson – a capital judge – informs us that the best he has seen for many years were shown at Birmingham in 1872.

Calls had been removed from the British Standards of 1901 (Waterfowl Standards were not published in 1886) and did not re-appear until 1971. There were probably very few Call Ducks to show in the intervening years, and they were confined to the Ornamental Class, along with the Pintails and Carolinas *(see* Fig 31). At that time, wildfowl were shown in the Ornamental Class at the shows. This probably contributed to the demise of the show Call because the Carolina and Mandarin are much more brightly coloured. Following a Poultry Club rule in 1958 banning the exhibition of all pheasant and jungle fowl, the British Waterfowl Association banned the exhibition of wildfowl (Hams, 2000).

Appleyard kept the Call in the 1930s and described it in his series of articles for *Poultry World*:

Fig 30 Grey Call Ducks. (Illustrated in Wright, 1874)

BELOW: *Fig 31 Bahama Ducks. The property of Mrs Sutcliffe. Drake: twice first at Dairy: Palace and Birmingham.* (The Feathered World, *25 August 1911)*

> You may fancy a really small, pure white duck with a wonderful voice, a very loud voice for so small a bird. I refer to the white Call Duck or White Decoy. When you set eyes on a really good shaped specimen – well it will please you: pure white with orange bill, legs and feet, very rounded in the skull with short bill, the body cobby, not long and slim, very short and tight in feather. They are really charming little ducks, and if unpinioned are great fliers.

Appleyard also had the Brown Decoy Duck, which he said was very attractive if it bred true to type and size. He described it as a Bantam Rouen, but said it was little kept. It was only by about 1946 that the Abbott Brothers began to list the Decoy again, for they had advertised it in *The Feathered World* between 1906 and 1910.

Vernon Jackson (personal communication) cannot remember Calls being shown at all in the 1930s and 1940s. When he judged the Ornamental Class at the Dairy Show in 1950, Appleyard's White Call was the only one. These birds were included in his Ixworth catalogue of the 1940s. A few years after Appleyard retired, he let John Hall have the best of his Calls in 1960–61. There were two drakes and three ducks, one of the drakes comparing favourably with the Dutch Calls imported a few years later by Jack Williams.

Dutch Imports in the 1960s

By 1964, Dutch Calls had arrived in Britain and were bred in their then standard colours by Jack Williams of Norfolk. John Hall remembers them as being quite different in type from Reginald Appleyard's. They were the true Dutch type, short in body, bill and leg. The imported colours then were White, Mallard, Pied, Silver, Blue Fawn and Black Bibbed. Later, Blue Fawn to Black Bibbed produced Blue Bibbed. Then Blue Bibbed to Blue Bibbed also produced Black and Lavender Bibbed.

Williams's birds were conventionally kept in their colour groups for breeding purposes and were not particularly easy to breed. However, in the second breeding season he was in hospital and unable to pen them in their breeding groups. The birds were kept with the wildfowl about three miles away from the house, and were left to get on with it in the mixed flock – and indeed they did. The results were marvellous; they bred wonderfully well. As Call breeders will know, this accidental move resulted in much improved fertility and vigour and, not only that, the standard colours were basically maintained.

In the 1970s, the Dutch breeder Cees Stapel bred Dutch Call Ducks in Suffolk. He described not only the White and Grey (Mallard) but introduced the term 'Appleyard'. He also had Pied, Fawn, and Black with a white bib. He considered that the Fawn was a colour mutation from the Grey. He described the females as:

> . . . a yellowish fawn colour with some blue shades on the wings, the drakes are of a similar colour to the grey drakes, but of a paler overall shade and the black in the tail is a bluish grey; also the head is a blue/grey.

He was describing what we now call the Blue Fawn, which was to prove the foundation stock for the Apricot, too. By 'Appleyard', Stapel meant what we now call Silver. The female, he said, was white with a lightish grey head and blue/green speculum. The drakes he compared to a Chiloe Wigeon in colour, with chestnut flanks, green head and some black in the tail. He found these birds prolific little layers, producing as many as 80–100 eggs in a season. He usually clipped their wings by the middle of March to keep the birds in their colour breeding pens. His later hatched birds were always smaller than the April/May hatches, a point with which Call breeders will agree.

John Hall remembers that interest in Calls was growing at that time. Several people kept them and he bred them himself in quite large numbers because they were popular; they all sold each season. There were also typed standards for the Calls, plus the Bali and Hookbill, when John Hall worked at Ixworth (Appleyard's establishment) in the 1950s. The Bali was fully recorded in the 1930 standards but the Call only got a one-word mention; perhaps Appleyard's Standard was from 1865. By 1954, the Bali was reduced to a paragraph, and the Call was classified as an 'ornamental' in white and brown.

Interest Grows

Interest was growing in the late 1970s from two further initiatives. Anne Terrell recalls going to a BWA open day at Jack Williams's in 1977 where she bought her first Call Ducks. She bought a trio of Blue Fawns, and the quality was quite unlike Calls she had seen before. The trio bred the colour sequence of Mallard/Blue Fawn/Apricot plus White and Silver. Since then she has persisted with this breed, particularly the colours, longer than anyone else in Britain apart from John Hall.

Also in the 1970s, The British Waterfowl Association began to hold its Annual Show and Sale at the Rare Breeds Show at Stoneleigh. The first of these was in 1976, where Stapel says there were only one or two Calls. But the quality that was to come was more interesting. About this time, American Call Ducks were imported by Christopher Marler in Mallard and White, and also bred by Fran (Alsagoff) Harrison and John Hall. Three pairs from Fran's White Ashfield strain were sold to a Cumbrian breeder, Brian Redhead. These American stock Calls began to turn up in larger numbers at the annual Show and Sale at Stoneleigh, where John Hall won the Champion Waterfowl Cup with one of his White drakes. The Cumbrian breeder Brian Bowes had also acquired some of the White American stock and it was for these that he became well known in the 1980s. Although, he says, the parent birds were quite big, they did breed Calls of a beautiful type, which were in high demand at Stoneleigh. Sometimes the American strain tended to breed birds with black feathers (a fault), but these tended to moult out and were not a problem.

Fran Harrison took a great interest in breeding Calls and it is largely due to her, John Hall and Brian Bowes that more birds were offered for sale each year. The Stoneleigh catalogues of 1983–85 show these three breeders offering the American stock for sale and coloured Calls as well. Whites were by far the most popular (twelve pairs in 1983) when the colours were restricted to only eleven pairs in standard Mallard, Pied, Blue Fawn and Silver.

Anne Terrell's interest in Calls was also taken up by Tony Penny about 1985 and Tony went on to win the Best Call Duck at the Stoneleigh Show and Sale for many of the following years. The numbers of Calls climbed to twenty-eight pairs of Whites and seventy-nine pairs of colours in 1992, indicating their popularity. Such was the demand for these birds, especially exhibition quality, that good pairs regularly fetched over £60, and the best over £200.

Call Ducks Today – Have They Changed?

Calls have never been so popular as in the 1990s. Instead of the two drakes on show at the Dairy Show in 1977, Calls frequently form over one-third of the total entry at many shows. Calls have certainly arrived because they are cheap to keep, easy to transport and show, but a real challenge to breed a winner. When the numbers have grown so much, there is opportunity for changes to take place in both type and colour.

Shape

Appleyard's Calls were rather long beaked and racey, but import of new stock from Holland in the 1960s and the USA in the 1970s, followed by the huge number of ducks bred in the 1980s, has meant that selection for the ideal Call type has radically changed the *average* bird. Selection from greater numbers has resulted in more Calls now being rounder and shorter beaked. Whilst top quality birds were there in Britain, in the Dutch stock of the 1960s and 1970s, they were still quite rare even in the early 1980s. However, it is now quite common to find several excellent birds at a show.

Colour

Whilst the birds multiplied, so too did the possibility of colour crosses. John remembers the Calls of the 1960s breeding in their standard colours even when they were all mixed up. They were White, Mallard, Blue Fawn and Silver. But when they are mixed and mixed several times, the colours eventually intermingle. Mallards develop colour faults, Blue Fawns acquire white flights and sometimes new colours are made. Black Bibbed can appear when Whites are crossed with Silvers. The Dark Silver arrived in the 1950s in Germany as a Mallard/White cross called the Trout (Sheraw, 1983), and this colour too also inevitably followed in Britain.

Call Duck colours were simple, or so it seemed in the early 1980s. Show schedules advertised only the standard colours though the Apricot, without a Poultry Club standard until 1997, did creep in. As Call Duck numbers have multiplied, so have the possible number of new colours, for Calls have frequently been mixed for breeding. This has produced new variants both here and in the USA where breeders are keen to get new strains.

Standard Colours

The descriptions that follow are not intended to be exhaustive descriptions of the Call colours. Systematic descriptions can be found in the *British Waterfowl Association Standards*, and a colour photograph is often as good a guide as any to the correct colours of the birds.

White Calls

White Calls are the favourites. Their clear, sharp outline makes them popular with wildfowl breeders as a contrast to the camouflaged teal. They constitute the largest Call class at a show because they can be exhibited for a longer season than the coloured Calls. There are no faded 'eclipse' feathers to contend with. The down side, so to speak, is that the bird has to be spotlessly clean and have no dark feathers. Nor should there be new, creamy white feathers to temporarily spoil the pure white. However, these yellow feathers often indicate a very

Fig 32 Line drawings of Calls redrawn from illustrations by different authors.
***1**. White Call Duck in Ives, 1947. The original photograph from the USA may date back to the 1920s.*
***2**. A pair of Appleyard's White Calls illustrated in his Ixworth sales leaflet of 1936. They have the correct body shape but lack the ideal head.*
***3**. Neither the American nor the British Calls are as short and cobby as the type illustrated by van Gink in 1932. This was, however, originally a line drawing and so the artist may have exaggerated the desirable features of the breed.*

fit duck, particularly when kept outside in winter in clean conditions.

The first Call Standard of 1865 said, 'bill bright, clear, unspotted yellow' but there must have been problems. In 1905 the American *Standard of Perfection* stated that any black in the bill colour of the drake was a disqualification, and this rule has remained. Unfortunately black can creep into the bill colour of several yellow-billed breeds and it is in the ducks that it is most difficult to eliminate. This is also true of the Call females, therefore specks of black in the bill, which encroach with age, are not regarded as a serious defect but there are now a few females that do not become mis-coloured with age.

Mallard, Blue Fawn and Apricot

Although these three standard colours look different from each other, they do exhibit the same colour pattern. All have a tidy bib, of dark claret in the Mallard drake lightening up to a mulberry bib in the Apricot. There should be none of these coloured

Fig 33 A top quality White Call Duck who won her class many times, as well as Best Call and Champion Duck, in the 1990s. As well as a short beak, she has a round head, cobby body and short legs.

feathers straying along the flanks, which should be stippled in French grey in the Mallard to light grey in the Apricot drake. The head colour is a typical green in the Mallard, changing to a charcoal blue (Blue Fawn) then pale grey in the lightest colour, the Apricot. The head colour is separated from the breast colour by an incomplete white neck ring.

Similarly, the ducks have a standard pattern based on the Wild Mallard. All three colours have the typical Mallard female face markings, showing a darker colour stripe passing through the eye, complemented by a paler stripe both above and below. The head feathers are marked with a darker graining and the body feathers are pencilled in the case of the Mallard, and laced with a lighter shade in the Blue Fawn and Apricot. Close inspection of the breast feathers does show pencilling is present in the Blue Fawn too.

Apricots are a particularly popular colour, but the Apricot ducks of 1999 are rather different from 1979. The 'originals' produced from the Blue Fawn were much closer to the American Pastel, having more blue-grey feathers on the back plumage; they could have been named the Apricot Blue (Anne Terrell). Over the last ten years the ducks have become brighter and more buff, probably because breeders have selected this colour.

To the unpractised eye, Mallard females are just little brown ducks and sound like that wild bird on the river. They are brown and do not compete for instant appeal with the alluringly named Apricots. Call specialists, however, love these brown ducks because the colour detail is a challenge and the shape of the bird can be exquisite. A well-coloured Mallard stands a good chance of getting Best Call.

Fig 35 Mallard Call female. This colour is less in demand than the White but is popular at the exhibitions. Getting the colour and markings right is an art. This duck is too light in the feathers at the outer edge of the tail.

Fig 34 Mallard Call drakes should mimic the Rouen for colour. They have a clean-cut claret bib, smart white collar, and grey feathers along the flank all the way to the dark under-tail cushion.

Fig 36 Apricot Call Duck showing the typical 'Wild Mallard' head markings. This colour of Call is very attractive and much in demand.

Finding good stock is difficult, however. If Mallard has been crossed with White, for example, the offspring can still turn out Mallard in the first cross. Such birds are useless for breeding pure colours, however, as they introduce colour faults in the next generation. Birds need to be bred in pure colours and selected for several generations, which is the usual practice for breeding Indian Runners, to ensure that colours breed true.

Silver Calls: Standard and Dark

These have been popular favourites for some time. The Silver is the standard colour but, over the last decade, the Dark Silver has cropped up too. The latter was definitely produced as a White/Mallard cross in Germany but probably not from a first cross.

Standard Silver Ducks have a white ground colour with a small amount of fawn colour mottling on the upper body and breast. The head is grained with dark markings and the duck should preferably have pale fawn feathers on the head and neck. The Silvers photographed at Stafford 1985 (Roberts and Roberts, 1986) show this as a typical characteristic. Both sexes have an iridescent blue speculum. The drake has the typical green head but his white collar completely encircles the neck and the mulberry bib is fringed with white on each feather. Unlike the typical Mallard, the brown bib feathers stray along the flanks.

Fig 38 Silver Call male at the same show. The mulberry feathers of the bib are fringed with white, and extend along the flanks. This is the key characteristic of the silver males. A very good type in a bird which often won his class, but was sometimes marked down for size.

Fig 37 Silver Call female at Devon and Cornwall Show, Wadebridge 1999. She shows the typical darker graining on the feathers of the crown and a good round head. The body feathers are a little pale but the speculum (blue colour of the secondary feathers) is intact.

The Dark Silver male is simply a darker version of the standard Silver. The way to identify them is in the belly colour. Dark Silvers are grey underneath whereas standard Silvers are white. In addition, Dark Silver males have a much more solid-coloured bib and more mulberry feathers along the flanks. There is no mistaking the dark female counterparts. Dark Silver females have been used for some time in the standard Silver breeding pen to maintain colour in the Silvers, which can become too light and even lose their bright blue speculum on the secondary feathers unless colour is re-introduced.

Pied Calls

As the name implies, they are white and coloured. The Mallard-coloured feathers of the head and body are in sharp contrast to the broad white collar and white flight feathers. The markings can be variable in extent but should be symmetrical. The pied gene, also seen in the Fawn-and-White Indian Runners, may also be bred with different colours. Pied can crop up in Silver and Apricot and other colours too, but only Mallard is a standard colour at present.

Bibbed Calls (Black, Blue, Lavender)

These pretty ducks should have a small bib of white

Fig 39 Pied Call male at the Shropshire and Mid Wales Exhibition 1998. (John Harrison)

Fig 40 A very nice Blue Bibbed Duck showing just two white flights and a neat bib. Quite frequently there is a white stripe behind the eye. This is very difficult to eliminate because it goes with the bib. Birds exhibited by Rosemary Sharpe in the 1990s have been excellent for type and colour in this difficult variant.

feathers on the lower neck and upper breast, in the form of an inverted heart shape. The earliest ones seem to have been black, as Black Bibbed is often the first bibbed colour to be produced from the Mallard gene pool, e.g. from White crossed with Silver. The most popular Bibbed now is the Blue. Blue Bibbed are simply known as Blue in the USA, where the colour was deliberately 'manufactured' during the 1970s from the introduction of a small Blue Swedish coloured drake (Sheraw, 1983); the ducks in fact follow the description of the Blue Swedish colour. However, in Britain, this bibbed gene emerged in the Calls direct.

Magpie Calls

These too have been developed relatively recently by breeders in Britain and the USA. They should mimic the standard-sized Magpie Duck in markings, having a black cap on the head and a heart-shaped mantle of black feathers on the back. Sheraw developed these in the 1970s in America. Like their larger Magpie look-alike, the Magpie Calls (and the Bibbed) have spotty rather than clear orange legs, and this is not considered a defect in the Calls. It goes with the colour markings, just as in the larger pied ducks like the Coloured-and-White Muscovy. Good magpie markings are difficult to produce in the Magpie Duck and even more so in the Magpie Call.

Fig 41 The Magpie Call is a very difficult colour to breed true, even more so than in the standard Magpie Duck. (Rosemary Sharpe's duck at the National, 1999)

New Colours

During the 1980s, non-standard colours of Call did turn up in several collections. Frank Mosford of Clwyd and Anne Terrell bred strikingly different pale lavender-blue Calls. There were pale 'Apricot' Ducks with a light ground colour, rather similar to the Butterscotch Duck of Darrel Sheraw figured in *Fancy Fowl* in the 1980s. These were called Silver Apricot in the 1990 Stoneleigh Rare Breeds Show and Sale catalogue. This name stuck because in the

1990s, the mixing of Apricots and Blue Fawns with a little silver also produced a darker variety of these 'Apricot Silver' and 'Blue Silver' Ducks, which were very noticeable because of their excellent type.

The gene pool of the domestic Mallard, and therefore of the Calls, does contain specific colours and patterns. Apart from the Mallard pattern, there are colours such as Dusky and Bibbed, as well as genes that produce the 'Silver' pattern (i.e. mulberry feathers along the drakes' flanks) in both the larger and smaller ducks.

These colours emerged to their fullest extent in the designer ducks – the twentieth century commercial crosses. Mrs Campbell wanted to produce a Buff Duck – but the insistent colour in her ducks was khaki. Cook had produced the Buff Orpington. In both of these breeds it was the dusky gene (inherited from the Indian Runners) that contributed to the new colours. As well as the buff colour, Cook also bred the Blue Bibbed and Black Bibbed in his Orpingtons. These 'Designer Ducks' released, much earlier, the colours being seen in Call breeding today.

Calls are now so popular that it may be possible to produce all the colours of these larger ducks on the Calls, though true Khaki tends to be elusive. The brown dilution from the Fawn Indian Runner, which also makes the Khaki, has been much more difficult to isolate.

The huge increase in numbers and colours of this breed has resulted, in the UK, in the formation of two Call Duck socieites. Numbers of Calls may be large enough to support more than the current nine standard colours. Popular new colours include the Yellow Belly in Holland, and Butterscotch from the USA.

Fig 42 Pair of Dark Apricot Silvers at home.

Black Calls

Tony Penny produced the Black Call from the Black Bibbed – with no admixture of Black East Indian to spoil the Call type. This is a distinctive colour and it will be interesting to see, in view of the history of the bigger black ducks, how long it takes to 'breed out' the white bib and brown feathers. Initially, the Calls are matt black (the Penny Black). Perhaps, like the Cayuga and Black East Indian, the birds will become more glossy all over as the white feathers of the bib are reduced.

Dusky Ducks

Prior to the 1990s, the existence of the 'dusky' gene had only been regarded as a fault in Calls although the effect of this gene was recognized in the bigger ducks. These Dusky Calls have probably been around in bird collections for a long time. John Hall remembers Jack Williams' collection in Norfolk in the 1960s containing not only standard Mallard Calls, but also the chocolate-coloured females of the dusky strain. The American Grey (Mallard) males, which were introduced into Britain in the1970s, also either lacked a bib or had an incomplete bib,

Fig 43 Dusky Mallard male bred and exhibited by Anthony Stanway at the Poultry Club National Show, 1999. The male lacks a white collar and mulberry bib, and the secondary feathers lack blue lights. The underwing coverts, like the Khaki Campbell, are also dusky. In a standard Mallard Call they are creamy white. As is usual with this colour, there is a small patch of white feathers under the chin; this seems to be the stamp of a Dusky at present. The lack of claret bib enhances the colour of the main body feathers, especially in the Dusky Blue male. Dusky males, too, often have a reduced sex-feather curl on the rump.

explaining why specific dusky faults in the Mallard were noted by Darrel Sheraw in America. In addition to the dusky look, John described these birds as looking rather like Laysan Teal, with light feathers around the eye as well as a lack of bib.

Until recently, the Dusky was observed but not remarked upon in Britain. However, using a systematic breeding programme, and close ringing the birds for specific identification, Anthony Stanway selectively bred this colour characteristic from his stock in the late 1980s. He produced not only the Dusky Mallard Call, but also the Dusky Blue and even Dusky Apricot between 1988 and 1999. The Dusky Mallard females were described as 'chocolate laced' because of their richer colour than the standard Mallard females. Duskies also turn up on the continent; a 'rare Dark Mallard' was figured in *Fancy Fowl* 1997 and a chocolate-laced Dusky female was imported in 1998.

The interesting point about these Dusky Ducks is that they are not a random mixture of colours. The colour is produced by a particular recessive gene. Once that gene is available it can also be used to produce other colours of Call in the recognized larger duck colours Buff Orpington and, perhaps, true Khaki. This has already been tried in the USA (as shown in the National Call Breeders *Call Duck Colour Poster*), perhaps because the potential and prevalence of the dusky gene in their Mallard Calls was recognized earlier with their breeders than ours.

Call Characteristics

Colour	Apricot Apricot Silver Bibbed Black Blue Fawn Chocolate Dark Silver Khaki Magpie Mallard Pied Silver White
Shape	Horizontal; large, round head; small, round body.
Size	Small: 1–1½lb (0.5–0.7kg)
Purpose	Ornamental

The Indian Runner

Indian Runners have the oldest 'pedigree' of all the domestic ducks. Although the Aylesbury beats the Runners to a written standard, the white table duck is a long way behind these egg-layers from the Far East in terms of an early start. It is said that there were representations of Indian Runners carved on the stones of the temple of Boeroe Boeddha in Java, over 1,000 years ago, when Britain had little more than the wild Mallard.

The Runners were recorded by Zollinger in Malaya (then a British colony) in 1851 as 'running nearly upright'. Wallace saw them in Lombok in 1856 and said that they 'walk erect, almost like Penguins'. They were similarly described by Darwin. The description of the 'penguin type' was replaced in 1877 by the term 'Indian Runner', possibly when the Fawn Runners were first shown at Dumfries. Any later references to the Penguin Ducks tend to get confused with a so-named designer duck (more heavy than the Runner but upright) which appeared briefly in the Poultry Club Standards of 1926.

During the 1850s, Lombok and Java were part of the old Dutch East Indies, so the Dutch as well as the British had colonial connections that allowed these productive birds to reach Europe by at least two routes. A Dutch naturalist stated in 1861 (in Brown, 1929) that the Runners are:

> . . . lean, with very few feathers, run nearly upright, and have a very long neck; they are principally reared on account of the eggs which are immediately salted, and for an article of food are much prized by the inhabitants.

Earlier than that, in the late sixteenth century, there are records showing van Houtman's ship carried a cargo of salted 'pinguin ducks'. Additionally an article of Dutch origin in *Avicultura* (1995) intimated alternative sources for the original imports, distinguishing between the less upright, Trout-coloured Ducks from Borneo, the Crested Runners of Bali and the slim-line Tegal Duck of Java. This variability in the Far East duck population probably accounts in part for the somewhat acrimonious debate over the proper Indian Runner 'type', which developed in Britain in the early twentieth century.

Why 'Indian'?

Prior to 1909 it was generally believed that the Indian Runner came from India. By then Matthew Smith had tried to find Runners in India and failed, as had the Indian Runner enthusiasts Digby and Miss

Wilson-Wilson. It seems that the name 'Indian' was applied because the ship's captain who brought the birds to Cumbria probably proceeded from India and traded with the 'Indies'. It was also thought that 'India or the Indies in those days included India proper, Burma, Malaya and a good slice of the East' (*The Feathered World*, 1925).

A good deal of mystery also surrounded the place of origin. Although Mr Walton imported true Runners in 1909, their place of origin was not publicly revealed until Dr Coutt's book on the Indian Runner was published in 1927. Similarly, the Misses Davidson and Chisholm also refused to divulge the origin of their Fawn Runners until 1929.

Productivity

The Runner Ducks were spectacularly successful layers compared with the indigenous types of the 1800s. Even after a prolonged sea voyage, the ducks would be back in lay after only two weeks. This productivity was probably the result of the rigorous selection that had gone on of necessity for hundreds of years in the Indonesian islands, which had a fairly high population density even in the 1800s. A Dutch writer (in Brown, 1929) found that, in the swampy areas on the south side of Java, the ducks were bred in their thousands. When they were 8–10 weeks old they were put in flocks of about 1,000 and began their long walk to be sold. Every day *en route* the young ducks foraged in the rice fields and even began to lay on this journey, which could take six months. At night, the ducks were penned in a portable rope fence corral. The eggs were collected and sold locally the next morning and also any birds that were unfit to carry on the journey. When Batavia (Jakarta) market was at last reached, the strongest long-distance walkers and best layers were there to be sold.

Duck culture was far more important in the Far East than in Europe. Powell (1993) considered that more than 70 per cent of the world's ducks were produced and consumed in South East Asia, where production tended to be less intensive that other types of poultry and depended largely upon peasant producers. In contrast with China's Pekin table duck, Indonesian production was for eggs, the green-shelled ones being preferred to white. An interesting point about the eggs, mentioned by earlier writers too, was the practice of partially incubating the eggs for eighteen days, then boiling them before serving them with salt. This delicacy known as *balut* is sold by street traders. The eggs were still incubated by traditional methods rather than electric incubators.

Runners in Cumbria

Britain's most frequent references to Runners in Victorian times come from Cumbria. Barrow-in-Furness built ships even then and the ports of the coalfield at Whitehaven and Maryport probably traded abroad. It was this trade that accounted for the import of the Far East ducks, though why they seem to have arrived here first rather than Bristol or London is a mystery. It is possible that the Runner did arrive first in the south of Britain (Weir, 1902):

> They bear a remarkable resemblance to some that Mr Cross had . . . in his rare collection on the lake in the Surrey Zoological gardens in about the thirties of the last century [1830s], . . . also about this time some were in the possession of the Earl of Derby at Knowsley.

A small pamphlet on the duck was written by a rather aptly named Mr Donald of Wigton in about 1890. He said that three ducks and a drake were brought from 'India' by a sea-captain to Whitehaven as early as 1840. A further consignment followed some years later, and he thought it likely that most Cumbrian Runner Ducks were related to these. The birds were widely distributed on the farms of Cumberland and Dumfriesshire but were, of course, intermingled with the local ducks for breeding, so that the original type became less pronounced. However, breeders became interested in the birds and, as early as 1877, Fawn Runners had a special class of their own at the Dumfries Show. In 1896 Miss Wilson-Wilson organized a display of twenty-one pairs at Kendal, when the Fawn-and-Whites appeared on exhibit for the first time.

Interest in the Runners grew. In 1898 a Mr Digby (Weir, 1902) also imported a trio and 'In 1909 a Mr Walton also secured birds from the Far East of the original type. For business reasons he and others refused to state whence these had been obtained' (Brown, 1929).

Standardization in 1901

> Towards the close of [1895] Miss Wilson-Wilson and Mr Henry Digby brought the variety under the auspices of the Waterfowl Club with the view of drawing up a Standard of Excellence, and then to induce a classification for them at the Poultry shows. (Weir, 1902)

In the first Poultry Club Standard in 1901, forty points were allocated for markings. This was not surprising because it was only the Fawn-and-White that was standardized; the markings are difficult to get correct. The Standard referred to both 'grey' and

Fig 44 Matthew Smith's Indian Runners. From the kindness of Mr Matthew Smith [a friend of Walton's] in lending me models, I have been able to have some photographs taken . . . . The side or centre view of the models is perhaps the most interesting, and, except that the photographer has not got them in absolute profile as directed, and that therefore the ducks look rather thicker through the body than in reality, it gives a very fair idea of the type and 'stance' of the bird that the Indian Runner Duck Club are aiming at . . . C was Walton's favourite Model, and A and B are recent. The present 'standard bred' is largely modelled on Walton's native imported duck of 1909, and owes most of its fine, thoroughbred, racy looks to these birds. (Coutts, 1923)

'fawn' but in both cases meant 'pied' birds, marked as in the Fawn-and-White (Wilson-Wilson, 1896).

In 1906, the Indian Runner Duck Club was set up with John Wilson of Appleby as the secretary. By 1907, its Standard of Excellence had been set up to revive the true type because, by then, the original Indian Runner had been crossed many times and was losing its original shape. J. Donald jun. (1905) castigated both Harrison Weir and Lewis Wright for depicting an 'Indian Runner' that was so 'absolutely at variance with their descriptive text'. So, in the 1907 standard, the largest share of the points was apportioned to the body shape and carriage 'the profile being suggestive of an old-fashioned soda-water bottle set at an angle of 60–70 degrees' (Powell-Owen, 1918). Jacob Thomlinson and J. W. Walton (1908) also championed the erect, slim-line type but despite all these efforts the argument, with much publicity in the press, was set to continue until 1930.

Harrison Weir's eventual (1910) comment on the standard is interesting. It perhaps indicated the conflict there was to be for the next twenty years over the correct Runner type:

> At the very outset, the fatal mistake was made of showing a decided preference for evenly marked plumage at the expense of true form and carriage. The idea then was to have nothing but clean-cut birds, with body markings somewhat similar to those of a Magpie pigeon, and had the club insisted upon type first, and markings second, all might have been well; but the craze for markings at any price soon brought disaster upon the breed. The pioneer breeders, disgusted at the ignorance of the writers and the decisions of the judges of that period, soon gave up showing, and kept their birds at home, breeding them to meet their own views.

The Utility Runner Duck Club

There was a body of utility breeders who had a different ideal from the Indian Runner Duck Club. They contended that the bird was too long and slim. E. A. Taylor (1918) of Hereford explained that he formed his Utility Club in 1915 to keep the strain that proved the most profitable for egg production. The narrow shoulders of exhibition birds would contract the 'crop'. Also, the perpendicular birds could not have the fullness under the tail, giving room between the pelvic bones, to lay large eggs. Taylor advertised his two types of White Indian Runners as the 'world's leading egg strains'. He said 'The hen will, I am sure, take second place to the duck as an egg producer'. Perhaps because the hen was more amenable to intensive housed production methods, this was not to be.

Exhibition Runners in the 1920s

Exhibition Runners have been bred for looks and the key features, when assessing a Runner, are the head and body shape, and the stance. Reginald Appleyard kept utility Runners (he became the Secretary of the Utility Indian Runner Club) but he also specialized in exhibition birds. In a series of articles in *The Feathered World* in 1925, he illustrated their key features. He stressed that the opinions expressed were simply his own, yet they have carried much

Buy from the Breeder who brought the White Runner to the front as a Laying Machine

AFTER many years of pedigree breeding and by a thorough system of egg-recording I have evolved the "JUST LAID" and "PALACE" strains of

WHITE RUNNER DUCKS

These represent the World's Leading Egg Strains and are distinctly unrelated to each other, so that customers can, if desired, cross them and maintain stamina. I supply the World's markets with eggs, ducklings, and stock, because everybody knows that my strains are the result of a thorough system of strain-breeding and egg-recording carried on for many seasons. Let me quote for your requirements in—

Eggs for Sitting, Day-old Ducklings and Stock Ducks

PHOTOGRAPHIC BROCHURE POST FREE

E. A. TAYLOR,

(Founder of the Utility Duck Club)

ROTHERWAS : HEREFORD

Fig 45 E. A. Taylor's advertisement for his utility Runners. (Powell-Owen, 1918)

weight over the years; many successful modern breeders still look to his ideal in a Runner.

In any Runner, Appleyard valued a dark, alert and bold eye placed high up in the skull. The beak, he said, should be wedge-shaped, neither 'dished' nor 'Roman nosed'. It should be of correct proportions, of a nice length, and fit imperceptibly into the head. This is difficult to describe, so he sketched what he meant.

With respect to the body and carriage, the standard demands a slim-line bird with narrow shoulders reminiscent of the classic 'hock bottle' shape. The neck of such a bird grades imperceptibly into the shoulders, which are just slightly flattened. The body needs to be tubular in shape with the legs set well down the body, well past the half-way mark, so the bird takes up an erect posture. When foraging or at rest, Runners often assume quite a forward, low stance but when they are alert or in a show-pen they

*Fig 46 One week's output. A pair of Taylor's 'Palace strain' of White Runners. (*Poultry World *December 1918, p.131)*

should stand at 80 degrees, or even vertically. Show birds should also be tall but in the pursuit of height the birds can become very long-necked and weedy and it is better to stick to the correct proportions of one-third neck to two-thirds body. This also allows a bird with a sufficiently round and full body to keep healthy. Weedy, inbred birds can drop dead from heart defects.

Correctly proportioned birds also stand better than long-necked individuals, which can go from one extreme to the other in stance. Some tuck their tail too far down between their legs and overbalance, whilst others turn up their tail like a Pekin.

There has always been a tendency for judges to go for the tallest Runner but Appleyard was quite clear about what he liked in the birds. It was certainly not just height. Writing in a *Feathered World Yearbook* he said:

> What is wanted is a Runner of medium size . . . a bird with a good, clean racy outline, and above all else a bird which can move and handle itself . . . a very long, slim, hollow backed bird is not a good Runner. What is wanted is a medium sized bird, free from any coarseness. Nice strong wedge-shaped bill free from any 'dish' in same, the bill fitting imperceptibly into a thin, refined head with eyes placed high in the skull.

Despite Appleyard's articles and models, the members of the Indian Runner Duck Club (Secretary: Mr Matthew Smith of Dumfries) still had to fight hard to champion their breed. The Misses Davidson and Chisholm, who imported the Bali Duck, obtained Runners in 1924 and 1926 from Java and Lombok (*The Feathered World*, 1929):

1. This sketch was taken from a Fawn Runner drake.
2. A White Duck:

   1 shows more length of bill than does 2, but both, to my eye, have bills which are balanced and in no way out of proportion to other parts of the head. Both have nice flat skulls, eye in the right place, good back skulls, nice necks.
3. This is a 'Roman nosed' drake:

   This Roman bill gives the bird a very striking and bold appearance.
4. Here I have shown the exact opposite in that we have a 'dished bill', a very common fault in many otherwise good or perfect birds.
5. Here we have a white, about which I can see nothing good, a throw back to Pekin . . . or some other white duck; poor, thin bill, rounded skull, thick neck and gullet, eye in the centre of the head.

Running one's eye over the five, one might take 2 as the best combined head and bill, but many would put 1 before 2. In any case, a really good duck, with perfect body and carriage, looks ordinary without a good head to set it off and make the finished Runner.

Quotes from Appleyard (*The Feathered World*, 1925).

Fig 47 The head of the Indian Runner Duck. (Appleyard, 1925)

Summary Key to Appleyard's Article (1925)

1. A good back line, direct from the back of the skull to the tip of the tail. No hollowness of the back, but a slight upward carriage of the tail.
2. Similar to 1 but the tail is carried straight down; Appleyard preferred 2 to 1.
3. An elegant and pleasing line from a duck that carried her neck slightly forward. The tail and stern are compact and short and are carried so as to throw the duck back into balance.
4. Neck not at such a forward angle; the tail is dropped again giving a very nice line.
5. Objectionable hollow where the back of the neck merges into the body. Together with the erect carriage and the dropped tail, this bird has the appearance of one that is trying to 'run after itself'.
6. This bird has a long stern and a hollow back. It lacks good carriage and may not be able to move without waddling.
7. This line is taken from a duck who always carries her tail between her legs, is almost too erect, yet is a very smart, slim bird with a straight out run.
8. This drake has very heavy feathering and a strong outline where the back of the neck merges into the body.

In summary, Appleyard preferred lines 1, 2, 3, 4. He thought 7 was an extreme line, 8 all right in some circumstances for breeding. Lines 5 and 6 were disliked.

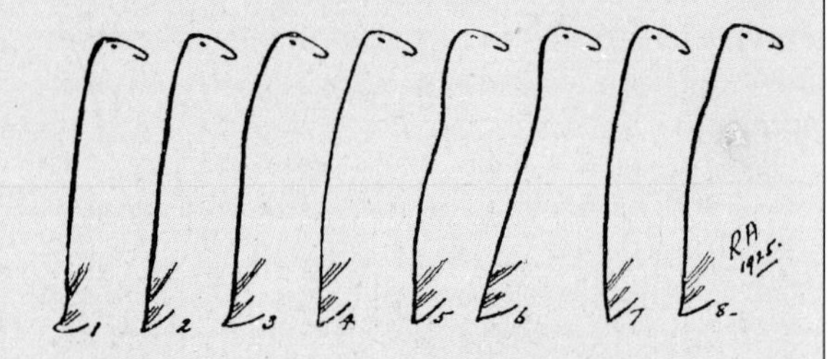

Fig 48 Reginald Appleyard's illustrations of the back line of the Indian Runner. (Appleyard, The Feathered World, *1925)*

> Our agent Mr Hanman declared them to be pure-blooded Runners, absolutely free from alien blood. We have now bred from this importation for four years, and have no hesitation in saying that it breeds true to type, nor does it show any sign of the eye stripe. . . .
>
> The island from which the Penguin type is obtainable is Lombok. We understand from Mr Hanman that it is now just about extinct, though it was plentiful in 1856, when Professor Wallace visited the island.

Yet in correspondence in *The Feathered World*, there was still a great deal of opposition to this 'penguin monstrosity type' by Ashe King, who conducted a public correspondence in opposition to those who promoted the true Runner. Coutts himself produced a book about the origin of the pure Indian Runner and, fortunately, the ideal of the Club survived. After thousands of years of evolution in the East Indies, there would have been little point in settling for the 'half-way' bird, advocated by the utility breeders.

Original Colours

The first Runners imported into Britain were probably Fawn, Fawn-and-White and even White. Coutts (1925) referred to White Runners being seen by British naturalists in Malaya amongst the flock of Fawns. White Runners, he went on to say, were introduced into Cumbria at the same time as the Fawn-and-Whites.

The first Fawn-and-Whites may not have been the same as the colour we are most familiar with today in the UK. The American Fawn-and-White (*Standard of Perfection* USA, 1898) is a bird coloured evenly in fawn. The 1901 UK standard also says that, in both sexes, 'the colour should be a uniform fawn or grey throughout the whole of the surface plumage . . .', i.e. it sounds like the colour

Figs 49 and 50 Miss Davidson and Miss Chisholm's Fawn Runners. Left: the 1924 importation. Right: That of 1926. (The Feathered World, *1929)*

Fig 51 Fawn-and-White Indian Runner: female's head showing graining on the brown feathers.

Fig 53 Fawn-and-White Indian Runner: male's head with a darker brown, solid cap. Ideally, there is a white line of feathers around the base of the bill and a white line around the eye, so that the cap of the skull is separated from the cheek marking (shown better in the drake).

Fig 52 Pair of Fawn-and-White Indian Runners. Males are taller than females. This colour of Runner tends to be thicker in the trunk (especially in older birds) than Trout, White and Fawn Runners.

Fig 54 White Indian Runners bred by Vernon Jackson. As well as being the top breeder of the exhibition Aylesbury, Vernon has also bred and exhibited White and Fawn Runners for many years and has supplied many breeders with their foundation stock. (Photograph by permission of Vernon Jackson)

Fig 55 Trout Indian Runners, originally bred on the continent from White crossed with Mallard, are very attractive. The ducks show slight variation in the amount of pencilling on the larger feathers, and in the amount of cream on the upper breast. This cream feature goes with the eye-stripes.

Fig 56 Head of a Trout Indian Runner Duck showing well-defined pale face and throat markings, and a darker stripe of feathers through the eye. Trouts have a particularly good head shape – a straight top-line to the bill, and eyes set well up in the skull.

first standardized in America (*see* page 37). Donald (quoted in Weir, 1910) referred to the:

> . . . American fad regarding the plumage of the duck. They fancied a drab or very pale fawn feathering in the coloured portions of the plumage of the female, and the feathers to be quite plain, without either lacing or pencilling. All brown and fawn ducks of the genuine breed . . . have the feathers marked in two shades of colour . . .

In 1923, it was eventually stated that 'the feathers of the duck have dark and somewhat brown centres with margins or fringe of a lighter shade', i.e. the feathers are 'pencilled', as stated in the beautifully written standard of 1926. It is this pencilled Fawn-and-White that is the standard colour in the UK today, the 'American' version only being added to our standards in 1997.

Apart from the original colours, there are now also Black, Chocolate (1930 standard) and Blue, which were all mentioned by Taylor.

On the continent, Schmidt (1989) states that after World War I, the Runner was bred to the Mallard pattern. In addition, the Trout was also obtained from France but also bred independently in Germany from a White/Mallard cross. In the 1970s, these two continental colours, plus Black and Fawn-and-White, were imported by Jacob Lory, and all eight colours were standardized in 1982.

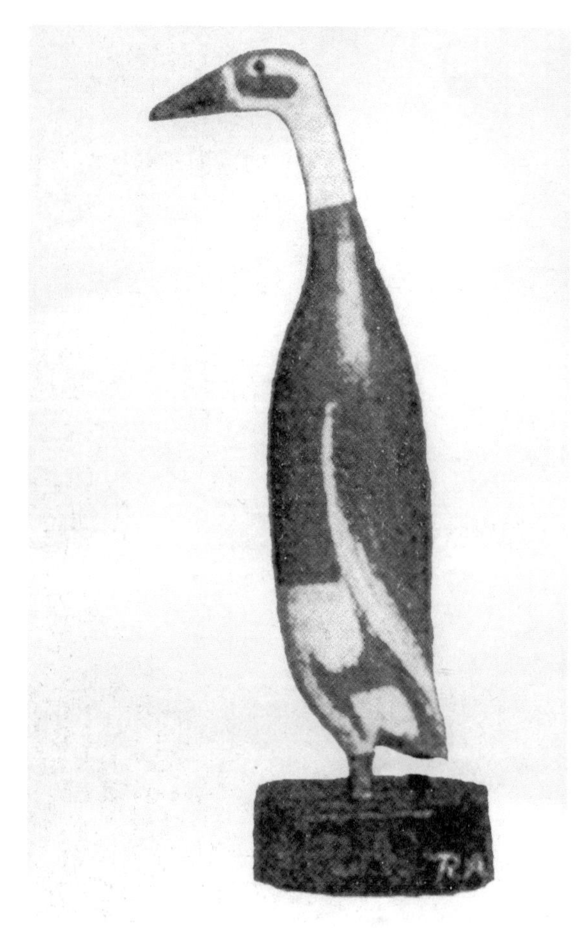

Fig 57 Model in wax by Mr R. Appleyard, Ixworth Suffolk, from his first, Special Reserve Waterfowl Cup Crystal Palace Duck 1924. Dam of his 1st Special Crystal Palace Drake, 1925. (The Feathered World Yearbook, *1926)*

In Britain, Runner breeders have, until recently, been much more conservative with colour breeding than in Calls. There has been no demand for Bibbed (viewed as a fault) but John Hall produced an attractive Apricot Runner from Fawn. Julian Burrell has also developed the same colour independently, as well as Lavender and Khaki Runners. Saxony Runners have been imported from Germany but have also been bred from Apricot Balis, which produced Blue Fawn, Apricot and Saxony colours quite independent of the German stock. In Germany, the 'Silber-wildfarbig' Runner almost reproduces the Abacot Ranger/Silver Call colour. In Australia, the Cinnamon is referred to as the same colour as the Silver Appleyard (H. Russell, personal communication). The Apricot Trout ('Saxony') and Silver are now standardized in the UK.

Indian Runner Characateristics

Colours	Apricot Trout ('Saxony') Black Blue Chocolate Fawn Fawn-and-White Mallard Silver Trout White
Shape	Very upright; narrow, wedge-shaped skull; straight bill; eyes high in the head; body long, narrow and cylindrical ('like a hock bottle'); straight line from head to tail tip; proportions of neck to body 1: 2
Size	Light: 3–5lb (1.4–2.3kg) Height (crown to tail tip): 20–26in (50–66cm)
Purpose	Egg-layer

Eggs or Exhibition

Although Indian Runners have the reputation of being egg-layers, they will not out-do a good Khaki Campbell strain in this capacity. Runners were, of course, instrumental in producing this brown utility duck, but even the earliest imported strains of Runners, judging from the Victorian descriptions, did not compete with the possible figures quoted for Campbells. Runners are a good egg-layer but there are big differences between the utility type and the exhibition birds.

Exhibition birds are not the best for utility. Tall ducks cannot be rushed about otherwise they will go 'off their legs'. I doubt if our British exhibition ducks are quite the same shape as the original duck from Lombok and Java. It is worth noting that it was because of the disparity in need between utility and exhibition that the Utility Indian Runner Duck Club was created and ran in parallel with the first Indian Runner Duck Club, 1906.

Runners do live up to their name and are extremely good foragers. Although they like water, they spend more of their time rooting about than heavy ducks and Calls, and are the ideal duck to rid large areas on a farm of parasites. They are more flighty than other breeds – though they cannot fly at all – and care has to be taken when they are new to a place. They are far more likely than other breeds to panic and run off. They need to be placed in a secure shed to start with before being let out into an enclosed run. When they have become used to the

*Fig 58 A Crested White Indian Runner Duck belonging to the Misses Davidson and Chisholm. (*The Feathered World, *1925)*

new place, or have noticed that there are other ducks, then they can be released with everything else as normal. Once they have become used to you, they arc calm with thcir ownєrs. Hand-reared, they are as tame as anything else.

Runners rarely sit but the occasional bird, if she has a nest she fancies, will have a go. The drakes are particularly bad with ducklings and will harry them and throw them about and even kill them. It is best to rear the ducklings yourself.

As with most domesticated animals, it is not always best 'to allow Nature to take its course'; birds have to be managed to get the best out of them. This is particularly true when the ducks are in lay; Runner drakes, along with Campbells and Harlequins, have a one-track mind. In the tougher utility ducks, repeated mating with the females is a nuisance but not often lethal. In Runners, however, the females get pushed over and trodden so frequently they become weak-legged and unable to walk. This situation can be pre-empted by allowing only one Runner drake per pen of ducks, if necessary.

Males that are surplus to requirements should be

Fig 59 A White Bali Duck bred and exhibited by Richard Sadler. She was Best Light Duck at Malvern and Stafford (Federation) Shows, 1999.

kept in a surplus drake pen, sold or culled. They can be disposed of as a flock of 'drakes only' for farm slug foraging or pond cleaning. Failing that, there is more meat than you think on a well-fed Runner. They are not worth plucking but the breast should be skinned and removed and will make a nice meal for two cooked in a good orange and honey sauce.

The Bali and Crested Ducks

The Bali is a crested duck. It looks rather like an Indian Runner with a top-knot. 'Crested Ducks', on the other hand have a rather average duck shape and size but a larger top-knot of feathers on the head. They can come in any colour but white is popular. They should weigh about 6lb (2.7kg) in the duck and 7lb (3kg) in the drake, though a larger Silver Appleyard type is exhibited. In addition, Crested Miniature Ducks now figure in the standards. These weigh 2–2½lb (0.9–1kg) and are simply scaled-down versions of the standard Crested.

The Bali

The Bali was originally imported from Malaya in 1925 by the Indian Runner enthusiasts, the Misses Davidson and Chisholm. It is an ancient breed, which takes its name from the island of Bali (east of Java) where Miss Chisholm believed it to be indigenous. Her Bali strongly resembled the Runner with its upright carriage, straight-out run and flat head with the eye placed high in the skull. The main difference was in the shape of the bill, where the very end was 'spoon-shaped'. The ducks were excellent layers of white eggs and therefore had a 'big capacity between the legs' and did not seem to be quite as racy as the Runner. The breed is therefore an old established one in Indonesia and, originally, was not a designer breed in the modern sense.

The Bali was standardized in 1930, but apart from this and another reference in *The Feathered World* (1925) when Coutts described his White Balis from the Chisholm stock, the breed seems to have virtually disappeared in Britain. In was briefly mentioned in the 1954 standard and again in the 1982 standard as a breed of duck that was 'more or less moribund if not extinct'.

During the 1980s, Tom Bartlett and John Hall became interested in breeding a crested Runner. With Tom it happened accidentally, it is alleged, when the White Runner got in the Crested pen in the breeding season. John Hall also, using several birds, produced a very attractive coloured line, which he bred in the 1990s to produce Apricot, Saxony-coloured and Blue Fawn variants. The Brown Bali Duck in the 1997 Poultry Club Standard was also produced from the same parents.

Although it was possible to breed attractive birds, many came without crests because Balis, like all crested ducks, have variable expressivity. John remembers, on a television programme about the birds set in Bali, that this was also the case. In the indigenous birds, the males all appeared to be crested, with a well-developed tuft, whilst in the females, several were plain headed. Perhaps the drakes, as is usual, were culled in greater numbers than the ducks. This might have resulted in only the more attractive male birds being retained for breeding.

Characteristics of the Breed

At present, White Balis are more numerous than coloured and this also seems to be the case in America. Sheraw describes only the white crested version in *Successful Duck and Goose Raising* (1975). He comments that the squarer shoulders and heavier body are characteristic of the breed; the body is never quite as slim as an exhibition Runner. Our standard also notes that it should not become too heavy and reminiscent of the Crested in shape. Unlike that of the standard Crested Duck, the Bali crest should not be too pronounced: a large crest usually accompanies thick shoulders. Also, the Bali has a fine Runner skull and therefore it can only manage a fairly small, round crest, which is set just to the rear of the crown. Weights at 5lb (2.3kg) in the male and 4lb (1.8kg) in the duck are similar to the Indian Runner.

Crested Ducks

Like the Hook Bill, the Crested Duck was also recorded in the seventeenth century in Europe and its long history again merits the breed's status as a Classic Duck. In this case, the early evidence is from Jan Steen, an artist from Holland, in his painting from around 1660 showing two Crested Ducks. Other Dutch painters, such as d'Hondecoeter, Dirk Wyntrack (1670), Pieter Casteels III (1723) and Aart Schouman (at the end of the eighteenth century) also depicted Crested ducks. However, in other countries there seems to be little evidence of the breed until Victorian times. It was mentioned in Browne's *The American Poultry Yard* in 1853 where it is referred to as:

> . . . the 'crested' or 'topknot' duck, a beautiful ornamental tame variety, which breeds early, lays freely, and hatches well. They occur pure white, black, or mixed with black and white.

Powell-Owen (1918) also asserted that the breed had existed in Scotland for over eighty years. It was a good table duck, both sexes weighing 7lb (3kg) or more. Despite this reference to the 1840s, the breed was not in the 1865 standard and Lewis Wright literally gave it a one-word mention in 1874.

Harrison Weir (1902) referred to it as the 'top-knotted duck' and said that it had been depicted in many old Dutch paintings in various forms and colours, some almost black whilst others were pure white. The bird had even become common in British farmyards, which is not surprising because if the Crested is not bred for exhibition qualities it is still a very good utility duck. It is a good forager and layer and of a reasonable size for the table. Weir, however, valued them as exhibition birds with a large crest covering the whole of the head. His beautiful illustration shows an excellent duck but he

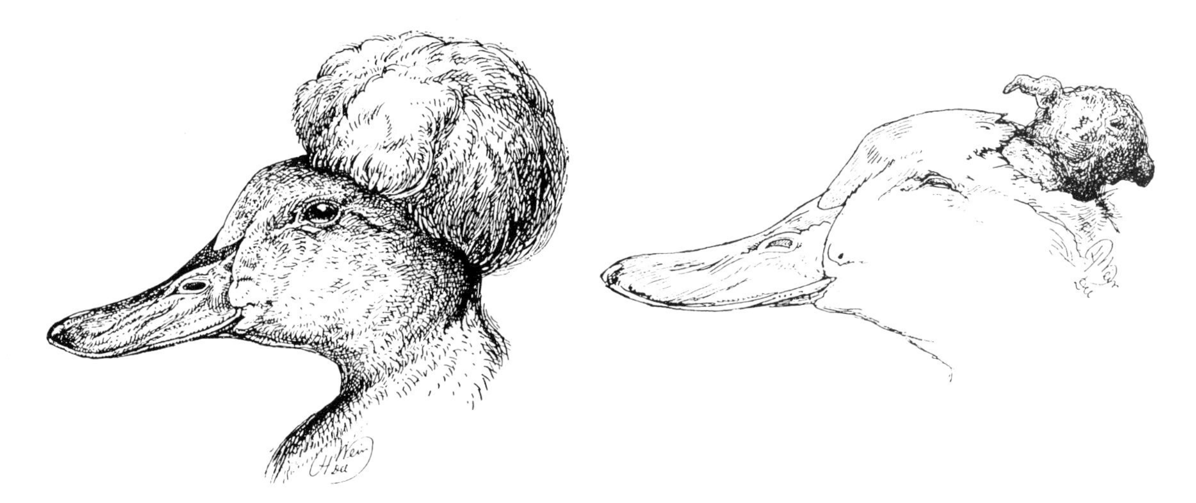

Fig 60 Duck with top knot. Owned by Harrison Weir. (Weir, 1902)

Fig 61 Head of duck showing fatty excrescence at the base of the skull. (Weir, 1902)

Fig 62 White Crested Ducks on a pool at Tom Bartlett's in 1999.

also drew the reality – that the attractive feathers were attached to an excrescence of fatty flesh. He commented that, by 1902, he feared that the breed was almost extinct. Nevertheless, it must have survived quite well for it was in our standards in 1910 in any colour and it had already been admitted to the American standard of 1874 in white.

The Crested Duck Club

Following Weir's influential publication in Britain in 1902, Crested Ducks became popular for a short time. In *The Feathered World* (1907) there was a proposal by R. Scott-Miller of Broomhouse near Glasgow, to form a Crested Duck Club:

> They are a useful and ornamental breed, as well as a very ancient one, and I think if fanciers of them would only combine and make it possible to form a club, we should see them rise greatly in popularity. I shall be glad to hear from any fancier willing to support the club.

Despite the efforts of the club, the Crested did not rise a great deal in popularity at the shows. A Crested Club Show (*The Feathered World*, November 1911) featured only two classes, four exhibitors and twelve entries.

Crested ducks disappeared from the show results in the 1920s and the duck was probably rarely seen in Britain in the 1930s. Appleyard did not feature the breed in his series of articles in *Poultry World* and he commented in his book *Ducks* (first published in 1937) that the Crested was by now uncommon, although from time to time a good one turned up at the shows. This was very much in contrast to the situation in the 1990s when the Crested became a popular breed.

A Revival of the Crested Duck

John Hall had always wanted Crested Ducks, but in the 1950s they were virtually unobtainable. However, Appleyard informed him of a breeder with a small flock in Norfolk and a few were obtained from this source. There were about ten birds, all with crests; all of them black with a white bib. The birds in the flock were old and had only produced one female the previous year. Using this young duck and a youngish drake, the two birds managed to produce one duckling from numerous eggs. This big, double-crested duck, was later paired to a Buff Orpington drake and then a large Silver Appleyard (from Appleyard's own strain) and these produced the future breeding stock. The original drake was never fertile again. Although John later found the odd White Crested Duck and Crested Mallards elsewhere, their crests were not very central and rather uneven. So the birds in his flock, which originated from the Norfolk duck, became his foundation stock.

John Hall's Crested Ducks must have been the starting point for many breeders, since few other people had these birds at all in the 1960s and 1970s. However, the Crested was to become more popular in the late 1980s and early 1990s when both John Hall and Stephanie Mansell produced them in a whole range of colours including Magpie (blue-and-white and black-and-white), black, khaki, blue, blue-fawn and pastel. Strangely, the pastel and apricot strains all relate back to a Hook Bill/White Crested cross, which made smoky blue ducks for John Hall in the 1970s.

Characteristics of the Breed

The Crested Duck weighs about 6lb (2.7kg) in the mature female and 7lb (3.2kg) in the male. They do tend to come in two types: a lighter weight version, a bit like a White Campbell; and the majority, a more solid type that feels heavier when you pick it up. Early references always stressed that this was a useful bird. The bill should be straight and fairly broad and the head set on a medium length, slightly curved neck. The breed is supposed to be a good forager, so it should have a slightly upright carriage of 35 degrees. The body is quite long and deep with a full, rounded breast. The strong wings usually just fail to meet over the rump. Most of the judging points are, however, for the head and crest, and many points are lost if the bird does not have a well-proportioned head and body.

Selecting Breeders

As pointed out by Oscar Grow, some characteristics associated with domesticated animals, such as the creeper gene in chickens, are inherited from lethal genes. In its homozygous form, where the offspring inherits the crest from both parents, the embryo will not hatch. This is the classic 'dead-in-shell'. Where only one of the parents passes on the crested allele, the heterozygous offspring should live and show signs of a crest (which may be quite variable in its expressivity). So, when two crested (heterozygous) parents mate that should produce 25 per cent dead-in-shell, 25 per cent plain-headed and 50 per cent crested. Conversely, a heterozygous crested parent with a plain-headed mate should produce 50 per cent crested and 50 per cent plain-headed offspring.

Some breeders like to hatch Crested Ducks because it is possible to tell very early on which are worth keeping for exhibition. The birds with

Fig 63 A blue fawn duck, an unusual colour in Crested, belonging to Stephanie Mansell. The symmetrical crest is well-placed on the centre of the head; the front of the crest is above the eye.

perfectly placed, symmetrical crests should be kept, and those with no crests or misplaced crests can be sold. Some of the birds will also behave in an odd way because they are crested, and these should not be kept either. They can be grown and fattened as table birds or, in the case of the females, they can be sold as layers. Crested Ducks lay well and can have high fertility and reasonable hatchability.

A point to think about when selecting breeders is that you must choose fit birds to achieve good results. As pointed out by Oscar Grow, selecting only for a perfect crest produces birds with poor conformation. Successful breeders of Crested Ducks always choose fit parent birds with no faults in the spine or defects in behaviour. A wry tail, a roach back (involving an upward curve in the spine), a kinked neck and a short back are all problems in this breed, if they are not selected against. It is wise to choose birds that are confident and move normally. The crest is *per se* a deformity itself and other deformities will go with it if they are not consciously avoided. A final tip in this respect comes from Steph Mansell, an experienced breeder of these birds: use unrelated drakes with the females in the breeding pen. In-breeding can emphasize unwanted genetic traits and it is best to avoid this tendency, especially with Crested.

Breeder birds should always have firmly set, single crests. A crest may grow from one or a number of roots and, if possible, birds of this latter kind should be avoided. The crest should also be set in the correct position on the head. The ideal one is placed so that the front is no further back than the eyes. It should not fall down the back of the neck, or be misplaced to one side. Birds that look like this are perfectly good for pets or utility birds, but are not for breeding. Note that whilst some breeders mate two exhibition birds, both with large crests, others have a different policy. It is generally acknowledged that better results follow when a large-crested bird is paired to one with a well-shaped but smaller, crest.

The Crested and the Bali

Is there a link between the original Balinese Duck and the Dutch Crested? Nobody knows. The crest itself has arisen naturally and has been selected for aesthetic purposes – this, to some extent, is what breeding has always been about with any domestic animal. The crest may have occurred more than once and in many different geographical areas. Odd birds with crests are produced in most breeds of domestic ducks and some species of wild fowl. Black-Necked Swans, Widgeon, Teal, Pintail, Mallard and Gadwall can all produce crested individuals.

An early record of a Crested Duck was unearthed in the 1368 estate rolls for Ramsey Abbey (Hams, 2000) where several ducks were described as being 'cirrar', which is thought to mean crested. Yet it is just possible that the Dutch connection with the Far East – Indonesia and Bali itself – may have brought the crested gene in the duck into Holland in the 1600s, popularizing the Crested as a result of the oriental spice trade.

When one looks at the dates, this does emerge as a distinct possibility. Van Houtman, who spied for the Dutch in Lisbon, collected information on the oriental routes that the Portuguese had established. The burghers of Amsterdam founded a 'Company of the Far Countries' and in 1583 entrusted van Houtman, with four armed ships, to follow the Portuguese spice route. They took a lengthy voyage, even visiting Brazil before eventually rounding the

Horn of Africa. They visited Madagascar before heading east, eventually reaching the Sunda Strait between Sumatra and Java, where they came into conflict with the Portuguese and the local inhabitants. They managed to make their way along the coast of Java, island-hopping to Bali. A further expedition set sail in 1595 and established a trading post in Ternate, the heart of the Moluccas.

Several other Dutch companies also set up rival expeditions until, in 1602, the Grand Pensioner, the supreme authority of the Netherlands, combined them as the Dutch East India Company. The Company planned and built a walled fortress in Java, which eventually housed 10,000 Europeans and Chinese. Founded in 1619, it was known as Batavia and, like Amsterdam, was constructed on a series of islets, making it a mosquito-ridden place. It was originally the site of a Javanese town, and the old buildings of Batavia still form a part of modern Jakarta (Taverner, 1972).

It is therefore possible that Crested Ducks, in the form of Crested Runners from Bali, were brought back from the East in the early 1600s. Kenneth Broekman visited the Maritime Museum in Amsterdam to research the records of the cargo of van Houtman's ship, the *Ysselstein*, which belonged to the *East Inde Compagnie*. He found that this ship transported hundreds of salted Penguin Ducks back to the Cape of Good Hope. He doubted if the duck's name was inherited from Indonesia. It is believed that the sailors named these ducks after their penguin-like appearance, for the same name was also used for the earliest Runners in Britain. The cargo also contained a large number of duck eggs but Broekman was unable to determine if these eggs were for breakfast or incubation.

The evidence is that the Dutch traded and travelled extensively in Indonesia, including Bali, and enough time had elapsed for the upright Bali Crested Runner to have crossed with the indigenous

Fig 64 Different types of crests in Miniature Crested. ***1.*** *A pretty, coloured duck with a large crest, possibly a split crest.* ***2.*** *The best bird of the four with a well-placed, symmetrical crest.* ***3.*** *A blue drake with a well-placed, single crest.* ***4.*** *A white duck with a good crest, but it interferes with the eyes, which is not a desirable characteristic.*

Dutch ducks to 'make' a European Crested Duck, as painted by Jan Steen in 1660. The Crested would have been used in Steen's picture because it was new and therefore a status symbol. It was the prerogative of the rich to have their portraits painted in that era, as well as to acquire rare and expensive products from the East.

Crested Miniatures and Crested Calls

In contrast to the standard Crested Duck, the Crested Miniature is definitely English and very recent. It is really not a 'Classic Duck' at all. It was created in the late 1980s and early 1990s from Crested stock originating from John Hall and also developed by Roy Sutcliffe. The breed arose from crested birds being crossed with Appleyard Bantam types and Calls. Coloured Miniatures in the Appleyard colours may have been the first ones produced, followed by the popular whites.

The breed itself was offered for sale by Roy Sutcliffe at the Stoneleigh Rare Breeds Show and Sale in 1992. It was first shown as a class at the BWA Club show at Malvern in 1994 and appeared in the Standards in 1997. In all points it is a miniature replica of the Crested Duck, weighing about 2½lb (1kg) in the drake and 2lb (0.9kg) in the duck. Consequently, any colour is permissible.

It is possible to produce crested varieties of other breeds. A Crested Black East Indian has arisen and the Germans had Crested Calls in 1962. Crested Calls also exist in Holland and the Pied variety is noted as being very plentiful (1997). Crested Calls are certainly not common and appear rather sporadically in the literature and there is perhaps a reason for this. The crest is created by a lethal gene, and small Calls are near the edge of viability anyway, even without a crest. John Hall remembers that when he tried to produce Crested Calls in the 1960s and 1970s that, each time they became Call Duck size, they tended to become infertile. He noted that a friend abroad also found the same result. The 1997 article on Calls in *Fancy Fowl* also reiterates that the crest is a lethal gene, and that crested Calls are 'a challenge to breed'.

The Pekin

The boat-people of China, living and trading from their sampans along the shores of rivers and harbours, have long been noted for their flocks of large, white ducks. Travellers have reported flocks of ducks mingling in the waters and padi fields during the hours of daylight but, at the evening signal, the birds would follow their owners' sampans back home again. These white ducks were probably distributed over quite a large area. They have been seen in Langshan on the Yangtse River and also in Swatow in South East China, where Woods related how hatching eggs were exported as far as Singapore and Bangkok. These eggs were partially incubated (to within a few days of hatching) so that the embryos were generating a considerable amount of heat. Arranged in a basket in layers three eggs deep, and insulated in soft paper, the eggs would hatch *en route* or on arrival. It was a time- and labour-saving operation in the warm climate, where the egg was the perfect receptacle for the journey.

Exports

Pekin Ducks, as they have come to be named, have been farmed since at least 2500BC in the Beijing area of China. They were immediately popular when they were exported to the West in the nineteenth century. They had aesthetic appeal: large, creamy-white, fluffy birds with the cuddly appearance of a teddy bear. They were healthy, hardy and, above all, meaty specimens. This latter quality made them suitable for crossing with white table ducks but, ironically, almost contributed to their failure to reach the New World as a breeding flock at the first attempt.

A Mr McGrath first saw them in Pekin (now Beijing) and was so taken with their size, similar to a

Crested and Bali Characteristics

	Crested	Bali
Colour	Any colour, as long as the markings are symmetrical	Any colour, as long as the markings are symmetrical
Shape	Reasonably upright (35–40 degrees); medium in body proportions; large, globular crest in the centre of the crown	Upright, as Indian Runner
Size	Light: 6–7lb (2.7–3.2kg)	Light: 4–5lb (1.8–2.3kg)
Purpose	Ornamental and dual purpose (eggs and table)	Egg layer, ornamental and exhibition

small breed of goose, that he resolved to export them to America. He hatched some eggs under broody hens in China. Fifteen ducklings, which were thought sufficiently mature to survive a sea voyage, were shipped from Shanghai in the charge of a Mr Palmer who was offered half of the survivors on reaching home. Six ducks and three drakes made the voyage of 124 days and were landed at New York in 1873; these should all have been the foundation stock for this new breed. Three ducks and two drakes were duly consigned to the McGrath family via people in the city, but they never made the final journey. These valuable ducks, despite having been described as 'dwarfed by the voyage, and not larger than good sized Aylesburys' had already been eaten!

Victorian Pekins

Fortunately Mr Palmer took better care of his own charges and the quartet soon recovered from the voyage. Having landed on the 13 March, they began to lay later that month; the three ducks managed 325 eggs by the end of July. In the same year, a Mr Keele in England also imported the breed and Brown also refers to a Mr Harvey who brought them directly from Pekin in 1872.

Palmer had originally seen the white ducks in Pekin and, from a distance, mistaken them for geese on account of their large size, long necks and large heads. However, they were said to be with some Barnacle Geese rather than large domestics and it is this comparison for size with the wild geese that probably accounts for the disparity in size and apparent 'dwarfing' of the birds when they reached New York, rather than their condition.

Poultry World (1874), quoted in Ives, described the birds much as they are today, but perhaps lighter in weight.

> The ducks are white with a yellowish tinge to the underpart of the feathers. Their wings are a little less than medium in length compared with other varieties. They make as little effort to fly as the Asiatic fowls and can as easily be kept in enclosures. Their beaks are yellow, their necks long, their legs short and red. The ducks are very large and uniform in size, weighing at four months old about twelve pounds to the pair. They appear to be very hardy, not minding severe weather.

Mason C. Weld wrote in 1874 that:

> In colour these ducks are identical with common yellow-billed white ducks, but in form, carriage and plumage they are quite different. The legs seem further back, or else seem shorter, so that it appears difficult for them to walk with their bodies horizontal, as the Aylesburys do, but they go with rather an upward inclination, so as to throw the centre of gravity back, rather penguin like. Their frames are very large, certainly larger than the largest Aylesburys. . . . The plumage has a peculiar cast under the feather, and a noticeable peculiarity in both sexes is a great mass of fluff and feather behind the rump. . . . Their heads are relatively large, and their necks very long. (from Brown, 1929)

Fig 65 Miss Croad's imported Pekin drake. (From Harrison Weir, 1902)

Miss Croad's imported Pekin drake had a lower carriage than the standard Pekin of today (Weir, 1902):

> When they were first imported they were not white, but of beautiful light colour, between a canary and a nankeen; and furthermore many of the drakes had manes, giving them a very novel and pretty appearance . . . But this novelty did not meet with the approbation of some who were 'set up' as judges of ducks . . . and so they continually passed these crested beauties until it was found there was nothing to be got by exhibiting them and thus this rare variation was bred out and it is difficult to find any of the old and true as imported.

Table Crosses

The great value of the Pekin at that time in the USA was its suitability as a table bird. Prior to 1873, the Cayuga had grown in popularity, but the black feathers were a distinct disadvantage for marketing the carcass – the dark fluff showed up. The Americans never seem to have been very impressed with the white English Aylesbury as a table duckling either, but they were taken with the hardiness of the Pekin. In the hands of breeders in the USA, the duck began to take on a different appearance from those developed in Europe. A line drawing of a drake exhibited in 1899 shows a more erect stance than the picture in the American *Standard of Perfection* for 1905 and the lower carriage of the American Pekin was confirmed by Brown who said in 1910:

> We have Pekins here, and prior to my visit [to America] I assumed that American Pekins were practically the same, but after conversation with Mr Sewell when he was in England, and seeing the American Pekin, I cannot fail to come to the conclusion that the American Pekin and the English Aylesbury are much nearer than we had supposed.

The imported ducks from the Chinese stock impressed the Americans with their 'thriftiness'. Reared for fecundity and economical meat production in China, they transferred these traits to America, where commercial duck breeders swore by their productivity. Rankin, known as the father of the Pekin Duck industry in America, hatched 15,000 ducks per year in the early 1900s.

> I have crossed the Pekin in every conceivable way with other breeds with an eye to securing a better market bird, but with unsatisfactory results. The birds either came out with weakened constitutions, or were longer maturing and had dark pin feathers or unsightly blotches on the skin. This experience has more than ever convinced me that there is nothing, as yet, in the shape of the duck that will supersede the Pekin as a market bird. (Rankin, 1910)

Starting in late November, his young ducks would lay about 140 eggs by August. The early eggs were infertile but by January were giving good hatches. Particular attention was paid to diet, meat protein being added to the early layers' rations. Later, green food and vegetables supplemented the hatching egg rations. Rankin made more money from his early ducklings and understood that well-fed birds on 'summer rations', even in winter, produced the most viable eggs and healthy ducklings.

Characteristics of the Breed

Colour

The British Standard for exhibition Pekins in the 1920s was virtually the same as the standard for today, except for the description of feather colour. Photographs of winning birds, even in the early 1900s, often show them as being less upright than today's, and sometimes to have had a very high crown and apparently dished bills – a fashion of the time and not the standard. The only main difference over the years in the description has been the colour, changed from 'creamy white' in 1901 to 'buff canary' in 1923.

A feature of the breed is the creamy-white feathers, especially when they are new and well oiled. The Americans certainly gave up striving for

Fig 66 Line drawings to show Pekins around 1900. 1. A fancy English Pekin drake. (Redrawn from The Reliable Poultry Journal, 1910) 2. The standard American Pekin. The standard drake already has a lower carriage than the 1899 winner. (Redrawn from the APA Standard, 1905) 3. First prize duck at the New York Show 1899. (Redrawn from The Reliable Poultry Journal, 1910)

'canary' when they settled for creamy-white, which was more appropriate, they felt, for a commercial duck. It may have been the impossibility of achieving the standard 'buff canary' that made the pure Pekin duck a rarity in Britain. The colour can be enhanced by feeding maize and colour-enhancing foods, but the natural colour is cream and this has been accepted as the standard in Britain too.

Shape and Carriage

In Britain, and particularly in Germany, exhibition breeders selected the upright type, which may have been closer to the original before the commercial breeders meddled with it in America. European breeders seem to have selected the creamier, fuller-feathered birds and in doing so selected a more upright stance. According to Selten (1996), the Germans obtained their first Pekins via France. The vertical posture was preserved and, he says, even accentuated by selective breeding. *Poultry World* figured a typical German Pekin in 1932 owned by Herr Schmeisser; the duck is not unlike the exhibition birds we have today.

Appleyard included a fairly upright Pekin in his book *Ducks: Breeding, Rearing, Management*, first published in 1937. However, Pekin Ducks rarely featured in *Poultry World* advertisements of that time and the ones photographed in *Poultry World* in the1950s were of a flatter utility type. Pekin strains that are currently in Britain originated from German imports, after the popularity of exhibition ducks and the search for new lines increased from the 1970s. This explains why the exhibition Pekins in Britain today are more erect. The head features of the high skull and short beak have also been exaggerated compared with the Victorian imports, and possibly the fluffy feathers. Consequently Pekins in Britain are now often referred to as 'German' to distinguish this upright type from the American, which has no standard in the UK.

Common faults in Pekins include (Fig 67):

1. not upright enough;
2. crooked back with raised chest;
3. flat shoulders;
4. body posture too upright;
5. body posture too horizontal and spread tail feathers;
6. double 'dewlap';
7. tail and wings carried too low;

Fig 67 Van Gink sketches of common faults in Pekins, courtesy of Avicultura. (See *text for details.)*

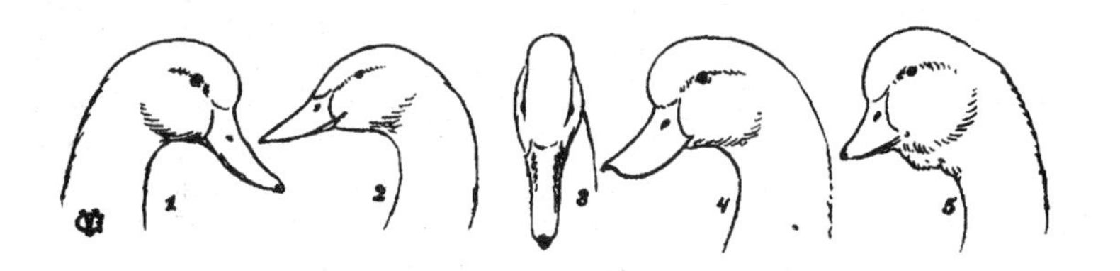

Fig 68 Van Gink sketches of the main head faults, courtesy of Avicultura. (See *text for details.)*

8. narrow hind quarters;
9. spread tail and wings carried too low;
10. type too small.

However, a keel between the legs, low on the body, is favoured by Selten and was mentioned in the British standards. Main head faults include (Fig 68):

1. curved beak;
2. thin, narrow beak with flat skull;
3. narrow, long forehead and beak;
4. mis-shaped beak;
5. fat head and throat dewlap.

Breeding Pekins

The pure Pekin is not a market bird today. Commercial duck producers select birds which are good layers and produce healthy hatching eggs; exhibition breeders produce birds for the show-pen. Whilst the two types of birds may not be mutually exclusive, one cannot expect show birds to be a 'utility' as well.

Young Pekins may start to lay in February or March, but the beginning of April is often the case for three to four-year-old birds. The laying season of April to July may produce up to 120 eggs, but fewer if the duck misses laying some days. Fertility will depend on the fitness of the birds and their diet, but this is not generally a problem in good conditions with grass for the birds for both eating and foraging in, and space for exercise. Hatchability will depend on incubator management. Pekin eggs are white, very large and round, and do not have an obvious air sac end. This presents two problems. Large, round eggs are more difficult to dehydrate to the correct weight for hatching than smaller typically 'egg-shaped' eggs. The duckling seems to have difficulty in deciding which is the air sac end too. Pekins are much more likely to 'pip' at the wrong end (i.e. the end without the air sac) than other breeds and will, in some cases, need help if they are not to suffocate.

Keeping the Birds

Part of the appeal of the Pekin is its soft appearance for, like all ducks, this breed has its own individual character. The Saxony is quiet and amenable. A Rouen will size you up and make a well-aimed strike with the wing if it does not want to be handled. Pekins have their own strategy – they nibble. All ducks, out of nervousness and self-protection, will often bite when picked up but a Pekin is more well-practised than most, and hangs on.

The profusion of feathers can present a management problem. These ducks do need clean conditions. The soft feathers get very dirty, much more so than any other white ducks, if clean bathing water is not available. This is exacerbated by the resting habits of the Pekin. It cannot gets its beak under the wing, so it sleeps flat out on the ground, and the mud. Although the breed is often described as having a long neck (and it is if viewed after it has been plucked) it appears to be short because of the fullness of the feathers. The neck is so packed out that it is impossible for the bird to sleep in the normal pose, beak under the wing. So they are very good at the 'dead duck' pose. Time and time again I have been caught out by a stretched-out duck, eyes closed to the world and apparently dead.

Partly because of their resting habits, and also because of their soft feathers, the Pekin does need more care than other ducks. Khaki Campbells can survive fairly happily in crowded, muddy conditions, partly because you will not see the dirt on the feathers but also because the tighter feathering will throw off the mud. In such conditions, Pekins will rapidly become very miserable indeed. Their soft feathers will pick up the mud, particularly if they do not have access to clean water to wash themselves and waterproof their fluffy feathers really well from

Fig 69 Pekins have loose, fluffy feathers, particularly on the head. They do need more care than other ducks to keep them clean and happy.

Fig 70 The Pekin's head is large and round and it has full feathers accentuating this. The beak is also short – it is the 'Call Duck' type of the heavy breeds.

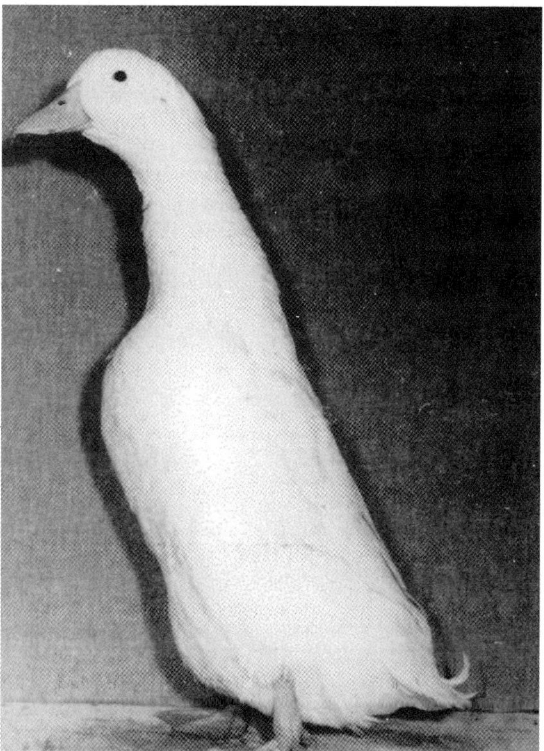

Fig 71 Pekin drake, Hannover 1962. (Rob Hilger)

*Fig 72 1st prize Pekin, Hannover 1932. (*Avicultura*)*

their preen gland. This is especially important as the birds get older, and in the spring as the feather quality begins to deteriorate.

Muddy conditions, especially with inadequate water for washing, also lead to eye complaints. Pekins are particularly prone to these because their abundant feathers tend to turn towards the rather deep-set eye. Mud seems to set up irritations too, so Pekins should frequently be checked for runny eyes. If such a condition is left, it can lead to the eye closing up and a very bad infection. Such a situation is best avoided by having clean water and, if an infection does occur, treating it early. It can often be checked by bathing the eye with a piece of cotton wool soaked with a very dilute solution of Milton fluid, and restricting the Pekin to a clean area with no mud and a few drops of Milton added to a bucket of drinking water. If the bird's wheat is in the bucket, the duck must wash its face to eat the wheat and the infection is cleared up more quickly. If this does not work, a tube of antibiotic ointment from the vet will do the trick. This, it will say, should be applied twice a day but will be promptly washed off by the duck. So the best course is to apply a liberal dose at night when the bird is shut up with no water. Also, look after Pekins carefully in hot weather. Despite being white, which should reflect the sun's rays, they suffer more than other ducks from the heat. They must have shade. This is particularly true of the ducklings, which will die of sunstroke whilst everything else seems perfectly healthy.

Finally, ducks suffer from fewer external parasites than hens, but the northern mite is very common. It is a red, blood-sucking mite, which can live all over the body of a neglected, sitting hen but seems to live mainly on the head and neck of ducks and geese. Perhaps the body temperature is too hot under the duck body-down for the mite. Unlike the red mite, which lives in the poultry shed, the northern mite must have the heat of the bird for survival, for if the bird dies the parasites soon die too. They can be found on the head region of slaughtered birds but they are fairly obvious in live, white birds if the feathers on the crown are searched. They can frequently be seen on white birds at shows, where the parasites emerge in the heat. It is extremely difficult to keep birds completely parasite-free. However, if the Pekins are scratching they do need a good powdering with a pesticide that breaks down in the environment. The fluffy Pekin feathers are more prone to harbouring mites and even lice compared with other ducks and irritation seems particularly bad in the spring if the birds are not treated. The parasite also exacerbates eye problems, so Pekins with runny eyes should also be checked for mites.

Despite these drawbacks, Pekins are great characters and well worth keeping. They also produce excellent commercial crosses and have been used in producing other strains, now recognized as breeds in their own right.

Pekin Characteristics

Colour	Cream plumage; orange bill
Shape	Almost upright; large, round head; short bill; body like a boat standing on its stern; heavily feathered
Size	Heavy: 8–9lb (3.6–4.1kg)
Purpose	Table bird

Black Ducks

Black Ducks come in two types: the heavy-weight Cayuga weighing in at 7–8lb (3–3.6kg) and the bantam-weight Black East Indian barely turning the scales at 2lb (0.9kg). Both breeds have black plumage with a brilliant beetle-green sheen, and the origin of one would seem to be linked to the other through the colour. Yet there is that huge difference in size to cope with, and also a difference in shape.

The origin of the Black Ducks is shrouded in mystery, for the truth is that nobody really knows if the Cayuga or East Indian came first, exactly where they came from or at what date they were first discovered or 'made'. Indeed, it may well have been quite a long evolutionary process that developed them, rather like the selective breeding that eventually produced the exhibition Rouen.

In the absence of concrete evidence, much of what was written by authoritative waterfowl enthusiasts was largely conjecture and contains conflicting information. Even the Victorians, who were closer to the original birds, misled later writers by using several names for what was finally called the Black East Indian. This breed was alleged to be from Labrador, and even Brazil and Buenos Aires. It finally ended up with a name from where it had never been seen and about as far as one could get from its true place of origin.

The Black East Indian

This is the Black Bantam Duck's name in Britain but the French still call it the 'Labrador Duck' and the Americans 'the Black East Indie'. It is said to have been first imported into Britain by the Earl of Derby and kept at Knowsley (Sheraw, 1990). It was named the 'Buenos Airean'. Some birds were also acquired by the Zoological Society of London from 'Buenos Ayres' (Nolan, 1850).

The breed was included in Wingfield and Johnson's *The Poultry Book* in 1853, where they referred to its plumage as 'deep glossy black':

> Metallic tints, varying with the light from green to gilded purple, decorate their garb of uniform velvet black, their bill and feet being of the same dark hue. . . . All who have kept these birds unite in expressing their constant annoyance at the appearance of more or less white feathers after their first moult. Mr Nolan, indeed, remarks, that they then frequently become entirely white.
>
> The first eggs laid by the Duck, in the beginning of the season, are frequently smeared with a dark, greasy matter, which causes them to appear of a

> slaty, and sometimes even of a black hue. . . . When six or eight eggs have been deposited, they gradually fade away to dull white. . . . The ducklings are somewhat difficult to rear, being very subject to cramp; this will not surprise us when we remember the warmer temperature of their Eastern Abode.

The above authors were very dubious about the duck's origin in Labrador, but they did seem to accept that it came from the East, as indeed did Harrison Weir (1902), after he had discounted Buenos Aires.

Buenos Aires may have been the original port of departure for the trading ship that carried the newly imported stock acquired by the London Zoo and so, true to the custom of the time, the birds were named accordingly. However, these Black Ducks had never been seen in South America. Harrison Weir questioned his son who had lived in the Argentine and who had travelled extensively, about the ducks. Weir was surprised to hear that the bird was unknown to his son, even though he knew all the other varieties of domestic ducks in that country. He concluded that the live cargo had therefore been picked up elsewhere. Weir also doubted that these were the first imports, 'for when I was at school in 1836 a boy brought some black and sooty duck eggs which he said were laid by the same black ducks that his father had got from abroad; doubtless they were from the same breed'.

So how did the name East Indian arise? The real ducks of the East Indies are the egg-laying breed, the Indian Runner. The original ducks of these were Fawn, or Fawn-and-White. Not only were they the wrong colour to develop the Black East Indian, but also the wrong shape as well. Sheraw amusingly remarks:

> Needless to say, the Malay Archipelago is not peopled with fanciers who have the leisure to create a miniature, ornamental black-plumaged duck suited in neither colour nor size for clean plucking and eating, whose unique value seems only to be its beautiful plumage, rather than its economic qualities.

He concludes that the name is one that took someone's fancy and cites several instances of names being made up for new colours and new crosses, which have nothing to do with the place of origin. Weir attributes their fanciful name to 'Messrs Baker of Chelsea who, after many importations, still kept the part of the "Indies" they got them from a profound secret'.

Although Sheraw had access to extensive American literature, he traced no specific early reference to the Black Bantam Ducks other than the British standard of 1865 and the American standard of 1874 when the Cayuga and Black East Indie 'were both admitted with numerous breeds which had been around for a century or more'.

The Cayuga

There is more information about the Cayuga. Ives was certain that the Cayuga was American: 'The

Fig 73 Black East Indians are smaller than Cayugas. They are about a quarter the size and weight of the heavy black duck. (John Hoyle, Whitchurch, Shropshire, 2000)

Fig 74 Cayuga drake at Blackbrook. Modern Cayugas have a green gloss over the whole of the feathers, and black legs and beak. As well as being larger, they are longer in the body than Black East Indians, and have a less upright carriage.

Cayuga duck is as distinctively American as the Plymouth Rock chicken, the Morgan horse or the Concord grape'. He quotes at length from an article in the Albany *Cultivator* published in 1851.

> We saw at Colonel Sherwood's in Auburn last summer, a singular variety of ducks, and on enquiry were told that they were obtained from Mr John S. Clark of Throopsville, Cayuga County. We were much interested in their appearance, especially in their resemblance to the wild black-duck (*Anas obscura*) that we wrote to Mr Clark to learn their history. In reply he said 'The ducks you enquire about have been bred distinct from any other variety at least twenty years. We obtained them some ten years since in Orange county and were told that they were originally descended from the wild black-duck, and from the great resemblance, I have no doubt the statement is true but cannot affirm this as a certainty.'
>
> The characteristics of this variety are, nearly a uniform colour (a little darker than the wild black-duck), good size, attaining the weight of eight pounds dressed at four months old, very quiet and very productive, one duck laying from 150 to 200 eggs in a season with proper care.

Fig 75 Cayugas at a pool at Tom Bartlett's in 1999.

Tegetmeir (1867) also quoted extensively from Bement, a well known transatlantic writer on poultry subjects:

> This bird derives its name from the lake on which it was supposed to have been first discovered. Of its origin, little is now known; it was quite common some fifty years ago in the barnyards in the vicinity of Boston. 'In the year 1812' says Dr Bachman, in a note addressed to Dr Audubon, 'I saw in Duchess County, in the State of New York, at the house of a miller, a fine flock of ducks to the number of a least thirty, which, from their peculiar appearance struck me as different from anything I had before seen among the varieties of tame duck. On inquiry, I was informed that three years before, a pair of these ducks had been captured on the mill pond. . . . The old males were more beautiful than any I have examined since, and yet domestication has produced no variation in their plumage.'

This quote does not actually say that these ducks were black, but American writers take this to be an early reference to American table ducks, probably the Cayuga. In addition to this reference, Jull (1930) refers to the breed originating from a pair of wild black ducks caught in Duchess County, New York State, as early as 1809:

> Whether they were descendants of the Wild Mallard, *Anas boschas*, or of the wild Black Duck, *Anas obscura* [*Anas rubripes*] is not known. Descendants of the Duchess County domesticated stock were taken to Orange County in 1840 and soon became widely distributed in Cayuga County, whence the name.

The importance of these references is that they pre-date the Black East Indian and refute one argument that the large duck was developed from the small one. Rather it might be that the Black East Indian was developed from the Cayuga. Perhaps more likely is Sheraw's hypothesis that the Black East Indian and the Cayuga were developed along similar lines for colour but from different source material.

An Alternative to the American Black Duck: the Mallard

Sheraw thinks it unlikely that black domestics have any connection with the wild American Black Duck, and there are good grounds for this too. Large black domestic ducks were known in both Britain and

America in the 1800s. In the USA, Browne (1853) mentions large black crested ducks and Weir (1902) found that even earlier, in the 1840s, a large black, white-fronted duck was much valued in the south of England. Crossed with the Aylesbury, they maintained their size and although some bred nearly white, and splashed black and white, some were even blacker than their parents.

The early American domestic Black Ducks often had some white feathers too. White on the breast was quoted as a common fault in the Black East Indian and Page in *The Cultivator* (1863) said that the Cayuga was black with a white collar on the neck or with white flecks on the neck and breast. The Cayugas were much longer and larger in the body than the Indians but not such an intense colour. 'Indeed by the Black East Indian . . . the Cayuga looks brown'. These birds could have arisen from ordinary domestics.

Black Bibbed can arise from Mallard and White crosses; this is how Black Bibbed Calls appear. Tony Penny used such Calls to breed Blacks, with no admixture of Black East Indian to spoil the type. As the bib is eliminated, it is likely that the Calls will lose their matt black appearance and acquire more of a gloss, like the Indian. If this is the result it establishes that the Mallard could have been the progenitor of the Black East Indian.

The process in reverse, perhaps, has now been described by Emmet (1999) from his closed flock of Black East Indian in Hong Kong. Twelve years of inbreeding (with no possibility of a wild Mallard cross) eventually resulted, not only in larger birds, but the hatching of a Mallard-coloured drake, complete with white collar and claret bib. His photographs also show a dusky female and dusky drake (no collar or bib). The dominant Mallard colour genes and the dusky are emerging again in the restricted gene pool.

Improvements in Colour

The Black East Indian was always an ornamental duck and, as Page's comment from *The Cultivator* (1863) shows, this breed was well ahead of the Cayuga in colour brilliance. According to Sheraw, this sheen on the feather was best developed in Britain. One export in 1871 of the Simpson Cayugas and Black East Indian went to a Mr Braikenridge of Chew Magna, Somerset. It was these birds that were used to intensify the colour of the Black East Indian and to make the breed available to other British enthusiasts. The 1874 edition of Lewis Wright's *The Illustrated Book of Poultry* contains a colour plate of a pair of Black East Indians and pair of Cayugas, which came from the Simpson stock. The birds are very similar in shape, the main difference being their size. The Indians are small in comparison, perhaps one-third or one-quarter the weight of the Cayugas.

The other difference between the breeds is the artist's use of green, this colour being more liberally applied to the back, head and wings of the Black East Indian. Note that in 1902 Weir remarked that the smaller the breed, the more brilliant the colour and sheen. Also, since prizes were awarded for colour and lustre, the size of the Black East Indian had been reduced as a consequence, which he regretted; he regarded the birds also for utility, i.e. for eating.

At that time, the Black East Indian had a dull olive bill; it is more difficult to decide on the colour of the Cayuga's but it was not slate black. The legs of the Indians cannot be seen in the painting but those of the Cayuga Duck are dull orange brown, as still described in the 1982 standard (though not in the American *Standard of Perfection* 1905 where they were preferably black).

In practice, over the years breeders have sought to improve not only the green sheen of both breeds of the birds but also the black of the bills, legs and feet. As a consequence, the British Standards now demand that the legs and feet of both Indians and Cayugas be as black as possible. The feather brilliance has always been a real challenge for breeders; birds with the most lustrous plumage are sometimes those without the desirable black bill.

Cayugas were originally developed as table birds and so their size had to be maintained. With breeders' efforts to maintain their weight and length of body it was probably quite difficult to develop the green sheen on the black to the same degree as the Black East Indian. Perhaps for this reason, the Cayugas used to merit more points in the show-pen for their body size and shape than their colour. The reverse was true for the Black East Indian. Today, the Cayuga has been developed for the show-pen to the same degree of colour brilliance as Black East Indian, but a common complaint is that the Cayugas are not large enough.

Breeding Stock in Black Ducks

Colour

Females can start to get white feathers in their first year, so it is desirable to select the breeders from those that acquire this characteristic later. All females will be turning white on the body feathers

Fig 76 Head of a young Black East Indian Duck. The eye is dark brown and the bill black except at the end where dibbling in the mud has removed the surface colour, and it is dull blue. There are no white feathers around the eye or under the chin.

by the time they are two or three, and by the time they are five they are nearly white, even though the drakes may be scarcely different from their first year. The males only develop a certain amount of white feathers around the eye, with sometimes a few white feathers on the neck and back.

A further tip from an experienced breeder is to select the breeding drakes from two-year-olds. As well as selecting birds with the fewest white feathers, it is at this stage that those with the best green gloss can be picked out. The inferior drakes will be showing more brown on their flank feathers.

On checking birds for green sheen rather than blue, good daylight is essential. When judging these black birds at shows, they are frequently taken to a day-lit area. The feathers covering the back – the scapulars – are easiest. The last places to acquire the gloss are the primaries, the stiff tail feathers and the under-tail.

In both breeds, the legs and webs should be as black as possible. Orange was tolerated in the Cayuga for longer than the Black East Indian but now top quality birds are definitely black-legged. Orange in the legs will come with age; it is more likely in the drakes than the ducks.

Today, exhibition birds in both Cayuga and Black East Indian have a black bill. Green in the bill is considered a fault; a lighter coloured bean (through wear and tear) is a fault to a lesser degree. Black East Indians have come a long way from those in the late 1800s and even early 1900s when the bill colour was a sort of yellow-olive, thinly washed over with black.

As for the eyes, the British standards used to say black but, if you look the ducks in the eye, the American Standards definitely got it right first time; the iris is dark brown.

Fig 77 A range of Black East Indian shapes. The shape of the body is quite difficult to describe in words and is better seen in the outline drawings taken from different types of bird. ***1.*** *This is the ideal shape. The head has a good crown, the neck is the correct length and thickness, and the compact body is well balanced. (The American Standard does say 'carriage nearly horizontal'. British judges prefer a slightly more upright carriage – 'carriage lively and slightly elevated'.)* ***2.*** *This bird has a head that is too flat for the typical Black East Indian. The neck and body are also too short.* ***3.*** *This bird is a type often seen in the show-pen but it is too racy; the duck has a reasonable head but is too long in the body. (Drawn from photographs in Sheraw)*

Size and Shape: Black East Indian

The breed is small and is classified as a Bantam Duck, weighing 2lb (0.9kg) in the male and 1½–1¾lb (0.7–0.8kg) in the female. According to *The Feathered World Yearbook* (1911):

> East Indias should be as small as possible, short in bill and body, [with] neat, round heads, and feet as black as possible, and free from white and brown feather.

Sheraw makes the point that the Black East Indian is in between the Call and the Mallard. The head is oval and refined but there should be a rise to the skull. The bill is moderately long, often 2in (5cm) but this is in proportion to the bird and not a rule. The bill is not dead-straight like a Runner, but nor should it be significantly 'dished'. A fault to watch out for is the elongated bill, which also flares upwards at the edges and exposes the sieves.

The neck is medium in length and the body 'moderately long' – not rounded as the Call, nor racy as the Mallard. Sheraw comments that the birds back in the 1920s in America (figured in Oscar Grow) dispel the belief that the older Black East Indians were large and racy. 'Black Prince' from *Feathered World Yearbook* (1911) also shows the compact type, rather at variance with the example from the 1982 standards. Nevertheless, this latter bird may well have been from the Black East Indian line imported from the USA in the 1970s. Fran Harrison notes that these American birds were smaller and a better type than ours. They had blacker legs and beak colour, and their eye was darker. However, the British birds may have been greener.

The birds carry their wings well up, so they fit snugly to the body; the breed is a good flier and can fly off if alarmed. This can even happen when they know their home, and they may still not return. When such birds are moved (on buying or selling) it is wise to clip them.

Fig 78 Cayuga drake: first at Dairy Show, 1903. Exhibited by Colonel R. S. Williamson. As in the case with many pure breeds of waterfowl and poultry, it is a few committed breeders who keep pure breeds going. Williamson continued exhibiting this breed into the 1920s. So also did Sykes, who took up the breed in 1897. (The Feathered World, *1905)*

Size and Shape – Cayuga

Cayugas are a different shape from Black East Indian. They are long in the body and the proportionately longer, upright neck is moderately arched. The legs are placed centrally and give the bird only a slightly elevated carriage. They should look a big, strong duck because the Cayuga was designed as a good table bird and a fair layer. They should weigh 8lb (3.6kg) in the drake and 7lb (3.2kg) in the duck.

At one time, the Cayuga did not have the feather brilliance of the Black East Indian, but over the years the green gloss of the feathers has been improved so that, today, the best Cayugas rival the Indian for colour. For this reason, the colour descriptions of the Black East Indian and the Cayuga are now the same and both breeds merit 30 points for colour.

Characteristics

	Black East Indian	**Cayuga**
Colour	Black-green plumage	Black-green plumage
Shape	Slightly elevated; compact	Almost horizontal; meaty
Size	Small: 1½–2lb (0.7–0.9kg)	Heavy: 7–8lb (3.2–3.6kg)
Purpose	Ornamental	Table bird

Footnote

Recent research (November 2000, personal communication) by Prof. Dr Wolfgang Rudolph has confirmed that the Black East Indian was indeed in the UK in the 1830s.

> Mr John Edwards (Zoological Society of London), with whom I am having an exchange of views on the history of breeds, recently came across an article published by the ZSL in March 1832. In the *Report on the Farm of the Zoological Society* it was stated 'A pair of Black Buenos Ayres Ducks also is in one of the ponds.' They were kept on the farm maintained by the ZSL at Kingston-on-Thames from about 1831 to 1834 . . . Mr Edwards is of the opinion that the ducks arrived in 1831 . . . It might be worth mentioning that the 13th Earl of Derby was elected President of the ZSL in the same year. Thus Sheraw might be right as to the foundation stock in Britain: the 13th Earl of Derby may have had a hand in it.

However, this was twenty years earlier than previously recorded (i.e. around 1850 in Sheraw, 1990).

2 The Designer Ducks

Plumage Colour in Designer Ducks

Why did it take until the twentieth century for duck breeders to become obsessed with creating new designs? It could be argued that the 'palette of colours' was not sufficiently diverse to produce many new breeds and colour variants until the introduction of the Indian Runners some time during the nineteenth century. When the basic stock included little more than the natural wild colour of the Rouen, the masking white of the Aylesbury and perhaps a few examples of black in the Cayuga and East Indian, there was little scope for experimentation. Yet by the opening decades of the twentieth century the situation had changed dramatically and a rush to design new breeds and new colours was clearly in vogue.

Unfortunately, the colour genes of feathers do not mix in the same way as pigment on an artist's palette. For example, mating a white bird to a coloured bird does not necessarily result in paler coloured offspring, as one might expect from the analogy of the mixed paints. A Rouen, say, crossed with a Pekin can produce youngsters that are almost black. Similarly, a Saxony crossed with a Pekin can throw Blue Bibbed, Blue-Fawn and other variants. Clearly it would be useful for the breeder to understand some of the rules that are in operation.

Most text-books on genetics are quite hard going, but there are notable exceptions. One such researcher is F. M. Lancaster, whose work on the inheritance of plumage colour in domestic ducks was made accessible in a number of stimulating articles in the BWA journal *Waterfowl* in the late 1970s. A collection of detailed papers had been originally published as *The Inheritance of Plumage Colour in the Common Duck (*Anas platyrhynchos *Linné)* in 1963, followed by a chapter called 'Mutations and major variants in domestic ducks' in *Poultry Breeding and Genetics* (1990). What ensues in this introduction is little more than a summary of some of the observations made in these articles, giving a rough outline of the main genetic ingredients of the most common designer ducks.

If we imagine the message of inheritance passed from each parent as a piece of tape on which there are special positions for particular details, then we have a very rough model of the chromosomes. At each position (or 'locus') there are a number of possible genetic options ('alleles'). There are simple options (like 'a or b') or triple options (like 'a or b or c'). Additionally, where one option is more powerful than the other, it is said to be 'dominant', whilst the other is 'recessive'.

Main Colour Variants

1. The most obvious position contains the information for **Wild Mallard colour** and markings. It is one of the triple 'a or b or c' series, which can be represented by the symbols M^R, M^+ and m^d. The + sign is used to show which allele occurs in the Wild Mallard.

   Mallard pattern (M^+): this is the wild form with typical olivaceous black ducklings, each with four yellow spots on the back, clear eye stripes and yellow throat and under-wing. The adults are most simply represented by the Rouen description found in the BWA standards.

					a or b or c	a or b or c	a or b	a or b				

Fig 79 Positions (loci) on chromosome.

*The biological reality is much more complicted than this, with numerous pairs of chromosomes, some of which are involved with the sexual differences and on some of which the genes for colour are mutually linked.

Dusky pattern (m^d): this pattern, recessive to the Wild Mallard pattern, is seen in the ducklings when the under surface is dark olive-grey; there are no yellow spots on the back and the under-wings have dark pigmentation. The adult drakes lack the white neck ring of the Wild Mallard, have little or no claret bib and the iridescent wing speculum is obscured. The under-wing is also pigmented. Ducks lack the typical eye stripes.

Restricted Mallard pattern (M^R): this is completely dominant over the other two patterns. The day-old ducklings have areas of dark pigment on the dorsal surface limited to the head and tail. The remainder is dull yellow with a dark undercolour. Adults are similar to the Wild Mallards but appear much whiter on the wing coverts. The feathers here are laced or tipped with white.

2. Another position has also a triple series of alleles, which interacts with the previous series to produce some of the most notable examples of designer breeds. Here the alleles, in order of dominance, are: Li^+, li and li^h.

Dark phase (Li^+): this phase is that of the Wild Mallard, detailed above. It allows full expression of the three 'Mallard' alleles and is top dominant in its own series.

Light phase (li): light phase day-olds can be recognized by white or pale yellow on the ventral surfaces and by single rather than double eye stripes. Adults are lighter in juvenile, female and eclipse plumage. Lower breast and abdominal areas are pure white rather than brown in the dark phase. Adult drakes have the claret bib extended down the flanks; light grey body colour turns to white and the neck ring is broader and clearer than in the dark phase.

Harlequin phase (li^h): the ducklings are completely yellow except for a faint smokiness on the head and tail. Adult females are even lighter than the light phase, owing to paler individual feathers and larger solid areas of white. The dorsal wing surfaces are mainly white whereas the light phases are coloured. The males can often be classified according to the further extension of dark claret from the breast to the flanks and by a clearer neck ring.

Interactions between the pattern and the phases are shown in the table below (based on Lancaster).

3. **Fully extended black (E)** is an autosomal gene

Summary based on Lancaster

Combination	Restricted M^R	Mallard M^+	Dusky m^d
dark phase Li^+	mismarked Rouen Li^+ M^R	Rouen, Mallard Li^+ M^+	Khaki Campbell, Buff Orpington Li^+ m^d
light phase li	Silver Appleyard li M^R	Trout Runner li M^+	Fawn Runner li m^d
harlequin phase li^h	–	Hunter's Mallard li^h M^+	Welsh Harlequin Abacot Ranger li^h m^d

Blue and brown dilutions on E and e^+ plumage types

Dilution genotypes	Combined with E/E	Breed examples	Combined with e^+/e^+	Breed examples
Bl/bl^+ $D^+/(D^+)$	blue-grey	Blue Swedish,	black areas of the wild-type diluted to blue-grey	Blue Mallard Blue Rouen
Bl/Bl $D^+/(D^+)$	blue-splashed white	light segregates from above	similar to above but paler	Apricot Call Saxony
bl^+/bl^+ d/(d)	uniform chocolate	Chocolate Runner	black areas of the wild-type diluted to chocolate brown	Fawn Runner Khaki Campbell Brown Mallard
Bl/bl^+ d/(d)	lilac (light blue with pinkish shade)	–	buff	standard buff Orpington
Bl/Bl d/(d)	lilac-splashed white	–	pale buff	blond Buff Orpington

that is completely dominant to non-extended black (e^+), the recessive gene, which allows full expression of the wild-type colouring when it is homozygous (i.e. matched with a similar gene). E, on the other hand, allows black pigment to be laid down in all areas except those affected by the genes for white spotting and interferes with all genes at the M and Li loci. This gene is found typically in the Black Orpington, Cayuga and Black East Indian breeds.

4. The following is one of three dilution genes (Bl, d and bu). When present singly or together in various combinations against the background of E or e^+, they are responsible for many colour effects in domestic ducks. The first is the **blue dilution (Bl)**. Incompletely dominant and autosomal, its main effect is to dilute black pigment into blue-grey in heterozygotes (dissimilarly matched alleles at one locus). Homozygous (matched) alleles produce paler blue and almost white.
5. **Brown dilution (d)** is recessive and sex-linked, its main effect being to change black pigment to dark chocolate brown and to obscure the iridescent wing speculum.
6. **Buff dilution (bu)** appears in reciprocal matings between Buff Orpingtons and Khaki Campbells as a further sex-linked recessive dilution gene.
7. **Recessive white (c)** is an autosomal recessive gene that can mask other colours and patterns. Present in the common domestic white breeds, this gene is said to be 'epistatic' when matched with the identical gene (i.e. homozygous). In other words, it conceals the effects of alleles in other loci.
8. **Dominant bib (S)** is a white marking on the neck and upper breast on a coloured background, e.g. the Blue Swedish or Blue Orpington. Lancaster (1963) found that the gene responsible behaved as an autosomal dominant and appeared to be closely linked with extended black (E).
9. **Recessive bib (b)** is a gene producing more constant shape and size of bib than the dominant form. Autosomal and completely recessive, it was found to determine an area of plumage roughly comparable to the drake's claret bib.
10. **White primaries (w)** is an autosomal recessive gene responsible for the typical Blue Swedish wing colour. The two or so outermost primaries are expected to be white, whilst the rest are coloured. There is a correlation between the number of white primaries and the extent of pale yellow markings (bill, feet and wing tips) on day-old ducklings.
11. **Runner pattern (R)** is similar to the above, in so far as it affects the outer primaries, but it is much more complicated. Found typically in the Fawn-and-White Indian Runner (and the Pied Call), this gene results in white markings in the following areas:

    upper neck region with narrow projections from the neck to the eye and up behind the bill;
    on the ventral surface of the lower abdomen;
    on the wings (primaries, secondaries and coverts).

12. **Magpie pattern.** Lancaster (1963) showed that the magpie pattern results when dominant bib (S) and the homozygous runner pattern are brought together, as in the Magpie Duck (genotype E/E S/S R/R). In this breed the head and neck are white except for a black cap on the top of the head. The back, rump, tail and rump cushion are also black, along with the wing coverts. The remaining plumage is white.

Other Traits

13. **White skin and bill (Y)**. In breeds like the Pekin and White Campbell the presence of xanthophylls in the diet causes yellow pigment to be laid down in the bill and skin. The Aylesbury, however, possesses a gene that prevents the deposition of yellow pigment in those areas. The Aylesbury has orange shanks and feet but the rest of the skin and bill are pink or white.
14. Another gene is the **Crest (Cr)**, which results in a raising of skin and feathers on the top of the bird's skull. Crested ducks are heterozygous for an incompletely dominant autosomal gene, which is lethal in the homozygous form. The homozygous offspring fail to hatch and often suffer from exencephaly, having parts missing from the cranium, and showing distortions of bill and spine. The gene responsible shows variable expressivity in both heterozygotes and homozygotes. In some of the heterozygotes the gene is not completely penetrant and the birds appear normal.

    Further genes include those causing hooked bills, egg shell colour, deformation of the embryos and hereditary tremor.

Sex-Linkage

In birds, the males have two relatively long (Z) chromosomes. The females have only one (Z) and a smaller sex chromosome (W). The female produces therefore two kinds of egg: a Z egg and a W egg. She can provide only one of the two chromosomes

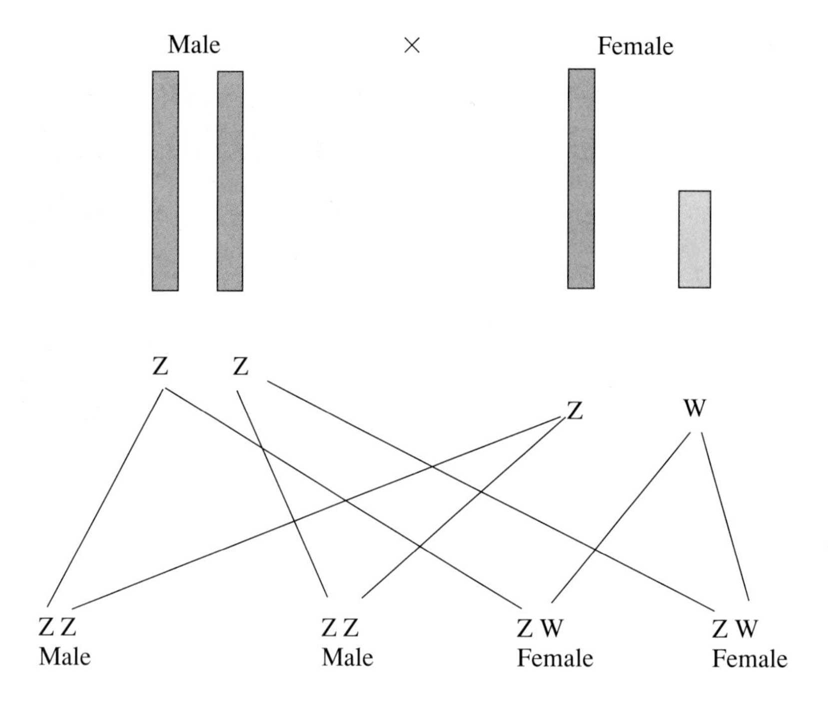

Fig 80

needed by the chick. The male provides the other, and since he only produces Z chromosomes *she* determines the sex of the offspring (*see* Fig 80).

In sex-linked genes, like brown dilution (d) and buff dilution (bu), which are carried only on the Z chromosomes, males have two (d d) and females have only one (d –). The hyphen, or minus sign, simply denotes that the gene is missing from the female sex chromosome.

The Campbell

No other designer duck has gained the success of the Khaki Campbell, and the reasons for this accomplishment we will try to establish later in the chapter. Simply it was the right product at the right time with the right name, but it was never the *only* Campbell: indeed there are four to carry the title.

Some time near the beginning of the twentieth century, Mrs Adele Campbell of Uley in Gloucestershire experimented crossing Indian Runner Ducks with Rouens. Just how pure these birds were is not known, but the breakthrough she made was to produce a bird with egg-laying capabilities similar to those of Runners but with a much more marketable body. The Campbell has always looked like a traditional table bird, whilst the Runner has much less credibility. Furthermore, the Campbell had all the benefits of a cross-breed: vitality, fertility and novelty. From the Original Campbell, which looked very similar to the Abacot Ranger, there developed the Khaki Campbell. Later emerged a white form. This was labelled simply the White Campbell. Finally a darker, more heavily pencilled variety entered the lists, to be standardized as the Dark Campbell. Only the Khaki had the fame and the catchy title.

From the beginning Mrs Campbell remained reticent about the exact 'recipe' for her birds. Paul Ives (1947) calls her 'a very secretive lady on that point', but she is not the first, and she is unlikely to be the last, inventor to conceal 'trade secrets'. In all honesty she may not have been able to vouch for the purity of her breeding stock. She was also adamant that her birds were not designed for the exhibition hall. They were egg-layers and, in the language of the contemporary idiom, 'utility' ducks. Reginald Appleyard (1934) made the following observations:

> First and foremost, the Khaki-Campbell was made with the idea of producing eggs, not meat. Mrs Campbell originated them in 1901, and they were made by mating a Rouen drake with a Fawn and White Runner duck, some wild-duck blood being added later. The Khaki-Campbell is a living egg machine, most strains producing perfect white or pearl-coloured eggs of lovely shell texture. Three hundred eggs per year is now quite common; not home tests, but official laying tests.

Fig 81 Mrs Campbell's ducks (Mark I). (Harrison Weir, 1902)

Appleyard studiously ignores the Mark I version of the Campbell and concentrates solely on the Mark II, the Khaki. He does, however, in the same article, cast doubts on the credibility of the simple parentage outlined above. He was a very experienced duck breeder himself, arguably the most celebrated in the twentieth century. His experience with his own cross-breeds will be discussed in later chapters. Simply the pied genes of the Fawn-and-White Runner are extremely persistent in the offspring, and the Khaki has no white markings. 'Considering the Mallard and Fawn and White blood – most dominant varieties – it is really a mystery how the Khaki has reached its present state of perfection in type [shape] and colour.' The only vestiges of white reported by early breeders were in the occurrence of a neck ring, on the male presumably, and a white patch on the throat of the female. Both of these were serious faults in the later Khaki version.

Fig 82 Campbell duck and drake. (From a photograph in Harrison Weir, 1902)

To return to the origins: the first birds produced and called Campbells were pictured in Harrison Weir's *Our Poultry* of 1902. There are two illustrations: one of a small flock and the other of a pair of birds. From such a limited sample and such a vague presentation it is still possible to see quite a lot of variability. Some of the birds have very wide neck rings and the females range from Rouen Clair to Abacot-type density of colour in the body markings. W. Powell-Owen, writing in 1918, ventures a description of these prototypes created by his friend, Mrs Campbell:

> The original Campbell males had dark green head and bills, grey backs, yellow legs, pale claret breasts, black sterns and a slight ring round the neck. The ducks had greyish-brown feathers pencilled with dark brown, brown heads, yellow legs and dark bills.

The creation of the name 'Campbell' for this breed of ducks is claimed by Joseph Pettifer, who wrote the 'Khaki-Campbell' chapter for *'The Feathered World' Yearbook* (1923). He too makes reference to the original bird:

> Perhaps it may not be generally known at the present day that at the outset the original strain was what might roughly be termed a grey duck. The Khaki colour was an afterthought, that came in during the course of perfecting and fixing the breed . . . It was in 1901 that they were first announced as a breed. In that year, when corresponding with me about them, she wrote: 'What shall I call them?' I replied 'Campbells,' and as such I first gave them that name in the Press. The prefix came afterwards, when the khaki colour came in.

The 1926 Standard

COLOUR
THE DRAKE

Bill green, the darker the better. Legs and feet dark orange
Plumage, – *Head, neck, stern and wing-bar* bronze, brown shade preferred to green shade.
Remainder of plumage an even shade of warm khaki

DUCK

Bill greenish-black. Legs and Feet as near the body colour as possible.
Plumage, – Khaki all over, the ground colour as even as possible. *Head* Plain khaki, a streak from the eyes considered a fault. *Back and wings* laced with lighter shade of khaki, lighter feathers in the wing-bar allowable.
Serious defects: Yellow bill; white bib; any deformity.

The Khaki Campbell

It took twenty-five years for the Khaki Campbell to find itself in The Poultry Club Standards. In the 'Addenda' to the 1926 version (above) are found preliminary standards for the extremely well-established 'Black East Indian Ducks', the 'Khaki Campbell Ducks', 'Magpie Ducks' and 'Penguin Ducks'.

It was not until 1930 that the first complete standard was published in which Mrs Campbell grudgingly allowed general characteristics and judging points to be drawn up. In both the initial and the present standards it is clear that great emphasis was placed on compromise. The Campbell was neither as tall, slim and upright as the Runner, nor was it large, squat and bulky like the conventional table bird. Utility and exhibition criteria had been carefully juggled.

The Khaki was expected to be quite small, unlike some of the show birds in recent years. The 1930 standard specified 4½lb (2kg) for 'birds in laying condition in their prime'.

The carriage was to be 'alert' and 'slightly upright', in other words somewhere between the extreme upright stance of the exhibition Runner and the horizontality of the Rouen or Aylesbury, 'the whole carriage not too erect, but not as low as to cause waddling'.

The 'type', or shape, of the Campbell frequently reinforced this compromise. The head should be 'refined in jaw and skull'; the bill should be 'well set in a straight line with the top of the skull'; the eyes should be alert, 'high in skull and prominent'. All of these are typical of the Indian Runner. On the other hand, the bird was to have a 'full face'; the body was to be 'deep, wide and compact'; 'activity and foraging power should be retained without loss of depth and width of body generally'.

In all, the Khaki Campbell was meant to be the perfect, all-round duck: a phenomenal egg-layer and forager, like its Runner ancestor, and a capable meat bird which was economical and easy to keep on the smallholding or commercial duck farm.

Colour

The American *Standard of Perfection* (1993) adds a note: 'Khaki colour can best be described as withered grass color, or as the color of khaki military uniform', but that depends on whose military uniform you choose. There is little uniformity in the shade. With ducks also, a single class of Campbell drakes can vary from pale fawn, not dissimilar to a Buff Orpington, to milk chocolate. Which shade is the correct one is open to speculation and deduction.

The *Shorter Oxford Dictionary* gives the definition as 'Dust-coloured; dull brownish yellow'. The word is said to come from Urdu ('*kaki*'), meaning 'dust-coloured', and fits in with the invention and early use of the drab uniform in the colonial outposts of British India at the time of the Indian Mutiny. Up to this time, most uniforms were brightly coloured – red or blue being very common. In a pitched battle it was vital to be recognized by comrades. Blocks of armed soldiers attacked blocks of armed soldiers. Visibility on a field clouded with gun-smoke was paramount; generals also needed to be able to see where their troops were deployed. Such was the state of military conflict at the beginning of the nineteenth century when Wellington was fighting Napoleon.

By the middle of the century, the state of affairs was very different, if not always so in the Crimea, certainly in the hills of India where guerrilla warfare was replacing the pitched battle. Concealed tribesmen could pick off red-coated soldiers with ease; so the apocryphal story goes that some of the soldiers deliberately soiled their uniform by daubing it with mud as a makeshift form of camouflage. How long it took officers to recognize the practicality of this insubordination is open to speculation. The credit is given to Sir Harry Burnett Lumsden for introducing khaki uniforms for the British colonial troops in India. It was widely used by the time of the Indian Mutiny (1857–58) and afterwards served as the official colour for British armies in India, and later it was adopted elsewhere and by other nations.

The original khaki was adapted to the terrain of light soils and dry vegetation. Hence it was made largely from light, dusty brown material, mainly cotton. Such material was later blended with wool and worsted fabrics and further adapted to suit climatic conditions in the South African War (1899–1902). This was all at the time of Mrs Campbell's experiments in duck breeding. Her son, on his return from the Boer War, is sometimes given the credit for naming the new breed for its similarity to his own uniform.

It was either luck or inspired marketing that allowed Adele Campbell to ride the wave of patriotic and military fervour. She produced the right bird at the right time when all the media and all the crowds almost literally worshipped their heroes in khaki. The word itself had acquired a kudos that would be the envy of later generations of advertising executives. Below the surface, however, is a less savoury element to the meaning of the word.

'Khaki' is not limited to Urdu. It is probably part of that body of language that was carried in migrations over the millennia and over the continents from north of India to the western outposts of Europe: the Indo-European group of languages. It is found in the Celtic languages, the Germanic languages, even in Spain and France, which were washed with invasions and migrations before and during the Empires of Greece and Rome. Because of its connotations it is likely that the word was not pushed out by the sophisticated and urbane replacements; it lingered in the language of baby-talk and vulgarisms. 'Khaki' means 'crap' or 'pooh' in most Western European languages. In modern Welsh, it is '*cachu*' (the 'u' being pronounced like a 'y' and the 'ch' as in 'Bach'), meaning 'to defecate'. It is '*caca*' in Spanish for 'dirt, filth and pooh'. The Germans refer to '*die Kacke*', 'shit' or 'crap'. Finally in France there is '*caca*'. Most appropriately of all, 'greenish yellow' is the shade referred to specifically by '*caca d'oie*', the French for 'goose shit'. The Khaki Campbell is a duck well designed for the rigours of the farmyard, but which shade of 'khaki' remains unclear.

Charles Roscoe, in his *Ducks and Duck Breeding* of 1941, raises the issue thus:

> The Khaki Campbells, when originated, were undoubtedly very near to the colour of Indian Khaki, which is lighter than the present army khaki. . . . At one of the last Crystal Palace Poultry Shows held, I sought out the Khaki Campbell classes, and found the winning birds were no longer khaki colour but a definite dark brown with very defined lacing on the breast, back and sides. They were not of the withered grass shade.

His assumptions are based on historical evidence. The use of olive tints was introduced to the khaki uniform during the First World War. The innovation was largely to reduce visibility of troops against the darker mud of Europe and the green foliage. The word now is roughly synonymous with any brown toned fabric used for military uniforms.

Markings

One of the most informative writers on the Khaki Campbell is Captain R. A. Long, who contributed a chapter in *Ducks* published by *The Feathered World* about 1926. His account is a gold-mine of information and, even when taking into consideration that this was a fairly early stage in the development of the breed, it is possible to see how deviations have crept into the common wisdom concerning the Khaki Campbell. I can do little better than summarize his key points:

- The underside of the wing should not be significantly lighter in colour than the rest of the bird's plumage.
- The 'wing bar' should not be defined with white and dark bars.
- There should be no light-coloured eye stripes in either sex. Where drakes have eye stripes in their eclipse (summer) or juvenile plumage, these birds should not be used for breeding.
- The underside of the drake's tail should be free from any grey or silver colouring.
- Also there should be no sign of claret on the drake's chest.
- The drake's bill should be pure dark green, not slaty blue-green (as indicated in the later standards). Long does, however, warn that pure green bills may easily turn greenish yellow at four to six months, whilst the bluer green does not have this tendency.
- The drake's neck colour should join the body at a definite and *level* line, not coming to a point.

- The pencilling on the back and wings of the female should be less heavy than that of a Rouen duck and the breast markings should be very faint, visible only at close quarters.
- Because of the 'pied' nature of the Campbells' ancestors, the Fawn-and-White Runners, special care should be taken to avoid patches of white, specifically on the neck of the female.
- Contrary to what later standards say, the head of the female should *not* be significantly darker than the rest of the body. Long calls this 'drake-headed'.
- Finally, the legs and bill of the female should not be too light. The legs should be the same shade as the body and the bill should be as dark as possible.

Captain Long draws attention to the pencilling on the body feathers of the Khaki duck. He is right in comparing the markings to those of the Rouen. The scapular feathers that cover the folded wings have 'concentric' lines of darker brown on a lighter ground. They look like double chevrons. A well-marked Rouen duck is a beautiful creature with golden chestnut pencilled in very dark brown. The patterns should be clear and luminous. The Khaki Campbell on the other hand should be subtle and restrained. Any pencilling should be much more delicate, especially in the contrast of shades. Blatant markings tend to be heavily penalized by experienced judges. Similarly, strong markings on the breast are looked at with disfavour not only by Long but also by Reginald Appleyard (1934):

> First, freedom from eye stripe; next, a soft warm shade of khaki throughout – definite lacing on top and thighs, not heavy Rouen lacing. On the chest I am of the firm opinion that there should be no lacing, but beautiful and definite ticking, just as if one had placed small ticks with the right-coloured pencil.

Fig 83 A pair of Khaki Campbells at Bill and Linda Bowen's (1999).

A final comment from Appleyard, about the drake: 'Go for drakes with bold, dark, alert eyes, pure dark green bills with no sign of greyness or yellow'.

Emphasis on Laying

It is one of the truisms of poultry breeding that 'a bird is what it looks like'. In other words, we tend to identify a breed by its outward appearance: if a bird looks like a Aylesbury, it *is* an Aylesbury, and so on. Until there is an accurate and convenient mapping of genes, we are forced to identify the breed by its phenotype, rather than genotype, and in most cases this has been adequate – not so, however, with the Khaki Campbell. In spite of all that has been written so far in this chapter, it is the utility characteristics that help to define the breed. Mrs Campbell herself resisted so long the attempts to standardize the Khaki; she knew that much would be lost if the breed became the sole possession of the exhibitors. Today we find it difficult to imagine just how crucial it was in the early years of the twentieth century for production characteristics. A flavour of this overwhelming preoccupation can be seen in Captain Long's enthusiasm for the results of the famous Bentley Laying Tests (1926):

> The popularity which Khaki Campbells achieved during the years 1921 to 1924 were undoubtedly due to the high records made at Laying Tests – notably the National at Bentley – both by flocks, pens and individuals. Official reports show the breed average of 160 Khakis to have been as high as 223 eggs in 48 weeks, while another year, 115 averaged 200 eggs in the same period, these being respectively 39 and 26 eggs more than the average of any other breed . . . Many individual ducks have gone over the 300 mark of course, and such records at Bentley include one of 346 eggs in 365 days, while the highest sequence officially recorded is 225 eggs in as many days to the credit of a Khaki Campbell.

Mrs Campbell's own attitudes are made equally explicit in a personal letter to C. A. House *(Ducks: Show and Utility*, *c.*1923). She also reveals interesting insights into the origins of the breed:

> The real beginning of the Khaki-Campbell was a great appetite on the part of my husband and son for roast duckling! To make ducklings one must have ducks' eggs, and I had just one duck, a fawn and

white Indian Runner, which laid 195 eggs in 197 days. She was the only duck in the yard, a rather poor specimen in appearance, and with no pedigree. However, I thought some good layers might be expected from her, but I wanted a little more size and mated her to a Rouen drake. Hence the lacing. The original Campbells were practically this cross, except that one season a Mallard drake was used. Then came the rage for Buff. Mr W. Cook was just bringing out his Buff Orpingtons, and I thought of getting a buff too, but failed. They would come Khaki. Just then the South African War was on, so I suppose it was patriotism. The foundation of the Khaki-Campbells was the original Campbell mated again to fawn and white Indian Runners. Then the trouble was to get rid of the white. This has now been quite overcome by careful breeding, and the laying power of the Khaki is greater than ever. Two ducks I have just heard of have beaten their ancestor hollow. One has laid a sequence of 225 eggs in as many days, and one 220. *There is no doubt it is easy to breed a duck looking like a Khaki-Campbell outwardly, but the laying inheritance is quite another thing* [Editors' italics].

A failure to heed the advice and methodology of these pioneers can find little better illustration than in the exhibition pens of the 1980s, a time when it was common to see pretty ducks in the farmyard that were light in weight and colour (conforming well to the original specifications), yet the shows were full of large, dark-coloured specimens that won many prizes but laid very few eggs.

The White Campbell

Developed in 1924, the white variety had to wait until the 1950s to make its entry in the standards. It was held to possess the same general characteristics as the other Campbells. It differed only in colour.

Captain F. S. Pardoe, who developed this variety, was aiming at a white breed with the same laying ability as the Khaki, yet with the potential of producing table cross-breeds for the commercial market. He wrote in 1936:

It should be possible to evolve a duck which will lay through the winter months and give the requisite 5–6lb duckling. If such a duck is evolved I have little doubt that the variety in which I am specially interested will play an important part in that evolution.

White variety:	
Plumage of	Pure white throughout both sexes
Bill, legs and webs	Orange
Eyes	Grey-blue

Laying Ability

According to *Poultry World* (13 April 1950):

At laying trials from 1932 to 1936 there were only seven White Campbell ducks entered, against from 55 to 144 of Khaki. In the 1936/7 Trials 175 Khakis averaged 225.71 eggs and the White (63 ducks) averaged 185.73 eggs. To gain a copper ring at Trials a Campbell must lay 220 eggs in 48 weeks. In the 1936/7 Trials 11.04 per cent of the 175 Khakis gained rings, and 8.33 per cent of the 63 whites. In the 1937/8 Trials of the 106 Khakis entered, 28.28 per cent gained the ring; in the 70 whites, 18.84 per cent. Nine Khakis laid over 300 eggs, up to 328 eggs; one White laid over 300 eggs, namely, 308.

The Dark Campbell

This version of the Campbell was first standardized in the 1950s (*see* table on page 72), the general characteristics being identical to the White and Khaki varieties.

A Mr H. R. S. Humphreys of Devon is given credit for the creation of this variety, which is said 'to make sex-linkage in ducks possible'. It was in the 3 September 1943 edition of *Poultry World* that the Dark Campbell was announced to the world in a discreet little article called 'Breeds Old and New'. During the Second World War, duck breeding had taken very much a back seat to chicken rearing, especially with the universal shortage of fresh eggs. It is surprising, therefore, to find such an unexpected upsurge in enthusiasm. The Dark Campbell was one of 'some forty breeds and varieties' of poultry represented at the annual display by The Poultry Club panel breeders at the London Zoo on 24–26 August 1943. Mr W. Powell-Owen was responsible for staging the show and he presented ribbons to C. Grange (Khaki Campbell Ducks) and H. R. S. Humphreys (auto-sexing breeds). The article gave the following details:

Mr H. R. S. Humphreys, of Lustleigh, Devon, exhibited auto-sexing Gold, also Silver, Welbars, as well as Dark Campbell ducks. The Welbar is the Barred Welsummer, and the sexes of the chicks, as hatched, can be told by differentiation in down colour.

By mating his dark ducks with Khaki Campbell drakes, Mr. Humphreys produces 'dark-down' day-old drakes and light or khaki ducks.

Dark Campbell

Drake's plumage: Head and neck: beetle-green. Shoulders, breast, underparts and flank light brown, each feather finely pencilled with dark grey-brown, gradually shading to grey at the stern close up to the vent, followed by beetle-green feathers with purplish tinge up to the tail coverts. Tail feathers dark grey-brown; coverts beetle-green with purplish tinge or reflection, also curled feathers in centre. Wing bow dark grey-brown laced with light brown; bar broad purplish green band, edged with light grey line on each side; flights and secondaries dark brown. Bill bluish-green with black bean-shaped mark at the tip. Eyes brown. legs and feet bright orange.

Duck's plumage: Head and neck dark brown. Shoulders, breast and flank light brown, each feather broadly pencilled with dark brown, becoming brown towards stern, with lighter outer lacing, followed by beetle-green feathers at the rump. Back and wing bow dark brown, outer laced with lighter brown. Wing bar as in drake, but less lustrous. Tail and wing feathers dark brown. Bill slatey-brown with black bean-shaped mark at the tip. Eyes brown. Legs and feet near body colour as possible.

Breed	**Male**	**Female**
Khaki Campbell	d/d brown dilution	d/- brown dilution
Dark Campbell	D^+/D^+ non-brown	$D^+/-$ non-brown

Male **Khaki** Campbell Female **Dark** Campbell

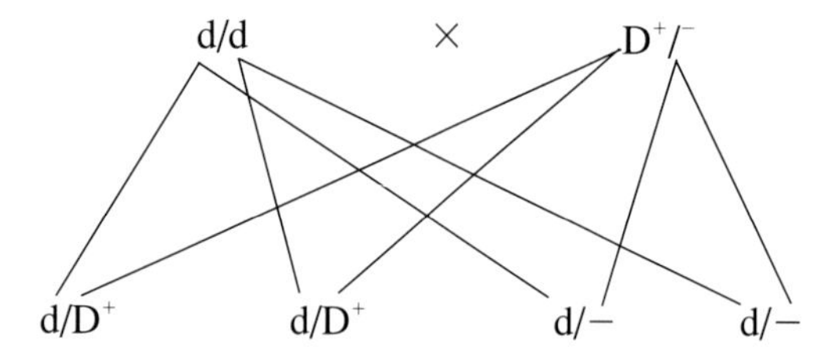

All **Dark** (looking) males All **Khaki** females

The School of Agriculture, Cambridge, discovered a sex-linked factor in ducks, affecting down colour and adult plumage.

This auto-sexing depends on the sex-linkage of the brown dilution gene (d), explained briefly in the introduction to this chapter. The Dark Campbell differs from the Khaki only in this one locus; all the other genes are identical. Because the brown gene is sex-linked it is carried by *both* of the chromosomes of the male bird and only *one* of the chromosomes of the female. Also, the brown dilution gene (d) is recessive to non-brown (D^+) (*see* table above).

If a Khaki drake (d/d) is mated with a Dark female ($D^+/^-$), the male offspring will carry one d gene and one D^+ gene, and, because D^+ is dominant, all the d/D^+ male ducklings will *look* like the Dark form, including having dark down. Of course they will be carrying recessive Khaki genes and not be pure, homozygous Dark Campbells. The female offspring, on the other hand, will have only the brown d gene,

Fig 84 A pair of Dark Campbells.

therefore be effectively pure Khakis, including the light down as female ducklings (*see* diagram, left).

It needs to be added that this auto-sexing only works with a Khaki *male* and a Dark *female*. If you try it the other way round (Dark male and Khaki female), *all* the F1 offspring will have dark down, and all eventually look like Dark Campbells.

The practical value of auto-sexing ducks has not been found to be great, compared with chickens. Very young ducklings are quite robust and easy to vent sex; thus colour-sexing has not had much impact in small-scale or commercial production.

Worldwide Fame

Mrs Adele Campbell's ducks, which started off in the orchard of a village house in Uley, Gloucestershire, have now made it to virtually all of the globe. Apart from its own marvellous egg-laying qualities, one of the reasons why the duck has become so famous and widespread is that commercial duck production using the Khaki has been run for decades by the famous Dutch duck company, Kortlang. The company originated in Holland, where the duck industry grew

Fig 85 A pair of Dark Campbells. (Illustrated in H. R. S. Humphreys, Poultry World, *1943)*

Campbell Characteristics

Colour		
Khaki		A shade of warm khaki all over, except for the bronze head of the drake and light pencilling on the duck
	Drake's bill	Dark green
	Duck's bill	Dark slate
Dark		Light brown with beetle-green head for the drake and dark pencilling on the duck. Purple-green speculum
	Drake's bill	Bluish green
	Duck's bill	Slate brown
White		White plumage; orange bill and legs
Shape		Slightly upright (35 degrees); medium in proportions
Size		Light: 4½–5½lb (2.0–2.5kg)
Purpose		Egg-layer

after the draining of the Zuider Zee. Displaced fishermen sought alternative occupations after the loss of their income and the Dutch Government became involved too. The Khaki's egg-laying potential was improved by adding the Tonsel strain and:

> . . . in 1939, Christian Kortlang came to England from Holland bringing 2,000 of these Khaki Campbells. All birds were numbered and their egg laying achievements recorded and today his farm in Ashford, Kent is the main source of commercial egg-laying Khaki Campbell ducks which are said to lay 340 eggs per year.

Exports have gone to India, Indonesia, Japan, Korea, Bangladesh, Taiwan and Central Africa. The company estimates that there could be about five million worldwide now – and all started by a doctor's wife from a back garden in Uley, Gloucestershire.

The Orpington

The Orpingtons were similarly the results of experimentation to produce an 'all-purpose duck', one that would lay a phenomenal number of eggs and still be competitive as a commercial table bird. I use the plural here simply because there were several different varieties emerging over time, a bit like the Campbells. There was the famous Buff Orpington, which vied with the Khaki Campbell for the most esteem as a designer duck; then there was the Blue, in all-blue and bibbed forms; finally there were Black, White and Chocolate, all surfacing in the first half of the twentieth century.

Letter from A.C. Gilbert, *The Feathered World*, May 1908

Madam – I enclose you a copy of the Buff Orpington duck standard, which I have just drawn up for this beautiful and most useful breed. I should be much obliged if you would give notice of it in your paper; also note that with other fanciers I am forming a club for buff ducks.

Buff Orpington Duck Standard

General Characteristics In Both Sexes

Head – Fine, racy, and oval in shape.
Neck – Fairly long and gracefully curved.
Bill – Moderate in length, straight in line from the skull.
Eye – Bold and full.
Body – Long, broad and deep.
Back – Broad and long.
Breast – Round and full.
Tail – Small, and rising gently, with the usual curled feathers in the drake.
Legs and feet – Strong in bone, well set apart, toes straight, connected by web.
General shape and carriage – Great length, broad, deep and squarely built, possessing an active appearance.
Size and weight – Drakes, 7lbs to 9lbs; ducks, 6lbs to 8lbs.
Plumage – Bright, with a nice gloss.
Colour
Bill – Yellow, with dark bean at the tip.
Head, neck, body and breast – A nice, even, rich shade of fawn buff throughout, the head and upper parts of neck in the drake to be at least two shades darker.
Eye – Brown iris, with blue pupil.
Legs and feet – A bright orange yellow.

Value of Points

Head and eye	10
Carriage and shape	25
Size	25
Colour	30
Legs and feet	10
A PERFECT BIRD TO COUNT	100

Serious defects for which birds must be passed – Twisted wings, wry tail, any other deformity. Lack of size. Colour other than stated.

'We now have two new varieties,' wrote W. Bygot in the *Feathered World Yearbook* of 1910 (he was referring to the showing season of 1909), ' . . . the Blue and the Buffs, both very useful for utility with regard to egg production and the table. The Indian Runner, too, is for the former purpose undoubtedly the best laying duck we have.' The Black and Chocolate Orpingtons were first shown at the 1920 Dairy Show.

Although beaten by the Campbells in laying tests and subsequent popularity, the Orpingtons were first in the race towards standardization. They were already in production by the turn of the century. William Cook was advertising not only the chicken of the same name but the Buff Orpington duck as well. He put a regular small advert in *The Feathered World* similar to the one illustrated for 19 January 1900.

William Cook and Sons (originators of the Orpington Fowls and Ducks), Breeders and successful Exhibitors of champion Rouens, Aylesburys, Indian Runners, Pekins, blue Orpingtons, and buff Orpington Ducks; several hundreds always in stock from 7s. up to £5 5s. each; pens carefully mated at moderate prices; birds on approval. Hints on Poultry and Duck Keeping, and full price list, post free. Only address, Orpington House, St Mary Cray, Kent.

Evidence of early standardization appears in a letter also in *The Feathered World*, of 1 May 1908, from a Mr Art. C. Gilbert, Swanley Poultry Farm, Willington, Kent (*see* box, *top*). It claims to be the first publication of a proposed standard for the Buff Orpington.

Mr Gilbert's format was little changed when the Standard was published by the Poultry Club in 1910, and again in 1922 when it included the Blue Orpington.

The Buff Orpington

The editor of the British 1954 standards introduces the breed with one or two disparaging remarks, which seem unjust in the light of the bird's later popularity especially in the 1970s and 1980s. He says:

> Its introduction followed that of the Khaki Campbell, and it has been said that the originator was trying to make a strain of the Khaki duck. At one time very popular for its high laying qualities combined with table properties and also its beauty of plumage and colouring, it has lost ground of late years, but a number of strains still exist.

He is right in so far as the Orpington never achieved the *commercial* success of the Campbell. Its ratings in the laying trials were high, but not as high as the Campbells and Indian Runners. It was as show birds and dual-purpose ducks that the Orpingtons managed to hold their ground. The plumage and

The Blue

Plumage an even shade, the darker the better, with a touch of white on the breast, the head and upper part of the drake's neck at least two shades darker than his body colour.

colour can certainly be quite stunning. He is wrong, however, in assuming the Orpington was a failed attempt at a Khaki Campbell. Mrs Campbell's letter (above) would suggest that the Campbell was indeed a failed attempt to emulate the Orpington (House, 1923):

> Then came the rage for Buff. Mr W. Cook was just bringing out his Buff Orpingtons, and I thought of getting a buff too, but failed. They would come Khaki.

In America the Khaki Campbell also achieved much greater fame, the two parts of its title being inextricably linked. The 'Buff Duck' is all that remains of the Orpington, and the American *Standard of Perfection* extant in 1947 gives it little more than a grudging few words. Paul Ives shows his disgust at this when he writes (1947):

> The sponsors of this fine duck and the committee on Standards which recommended its admission . . . must have had little imagination and even less interest when they recommended for the name just simply 'Buff'. It's like naming a breed of chickens 'White', a breed of cattle 'Yellow' or a family of horses 'Bay'. Small wonder the name has never attracted the attention of the public and that the breed has never forged ahead and flourished.

Ives's *Domestic Geese and Ducks* (1947) is a book full of useful information and brimming with sense.

I think he is right, but not totally fair. The word 'buff' did not have quite the advertising flare of 'khaki', especially at the beginning of the century, and more especially in America. Not only is the colour more subtle than khaki, so are the connotations. In theory the creators of the Buff Orpington should have been able to cash in on the military fervour just as Mrs Campbell had done. Orpington is in Kent. 'The Buffs' were soldiers from the former East Kent Regiment, its uniform facings being of that colour. The contemporary and parochial nature of the reference became eventually lost, whilst khaki gained currency worldwide.

'Buff' is described in *The Oxford Dictionary* as 'a yellowish beige colour (buff envelope)' and can refer specifically to a velvety dull-yellow ox-leather. The word itself relates to 'buffalo' (*boubalos*, Greek for antelope or wild ox). The Orpington, however, is more complex in colouring. It carries a definite reddish undertone. Later British standards (1930–97) describe both drake and duck as having:

> . . . plumage a rich even shade of deep red-buff throughout, free from lacing, barring and pencilling except that the head and neck [of the drake] are seal-brown with bright gloss, but complete absence of beetle-green, the seal-brown to terminate in a sharply defined line all the way round the neck. The rump [is] red-brown, as free as possible from 'blue'.

The bill and legs also carry elements of orange and red.

Breeding Problems

Hardly anyone will tell you, when you start breeding Buff Orpingtons, that these ducks do not breed true. They suffer from a genetic complication that also afflicts all the standard blue breeds of ducks, i.e. they carry a heterozygous, incompletely dominant characteristic. If we look at a representation of the bird's genotype we can see that the colour genes all match up in the two inherited strands of genes (the chromosomes) except for the blue dilution gene:

> m^d m^d [dusky] e^+ e^+ [non-extended black]
> d (d) [brown dilution (sex-linked)]
> bu (bu) [buff dilution (sex-linked)]
> Bl bl^+ [Blue dilution + non-dilution].

In fact there are *two* Buff Orpingtons:

> the dark coloured standard Buff Orpington:
> m^d m^d e^+ e^+ d (d) bu (bu) Bl bl^+
> the pale coloured non-standard version:
> m^d m^d e^+ e^+ d (d) bu (bu) Bl Bl.

M. Lancaster, in his article 'Revitalizing the Buff Orpington' (1980), explains it as follows:

> It will be noted that the blue dilution gene (Bl)* is only present in a single dose. This heterozygous (impure) constitution (Bl bl^+) is necessary for the birds to show the rich buff colour required by the breed standard. Homozygous (true breeding) birds (Bl Bl) tend to be too pale, particularly in the neck region, to comply with the standard requirements. Unfortunately the standard-type birds are not true breeding and when mated together produce a ratio of:
>
> 1 light buff (Bl Bl) : 2 standard (dark) buff (Bl bl^+) : 1 khaki (bl^+ bl^+)

> in the progeny. Breeders of Blue Orpingtons, where mismarked splashed white and black ducklings are produced, will appreciate the problem. This is a further factor affecting the popularity of the Buff Orpington.

*He uses the older symbol (G) in the original article.

Origin

Credit for all of the Orpingtons, including the famous varieties of chicken, goes to William Cook and his family. Cook used a similar recipe to most creators of 'designer ducks': he simply crossed Indian Runners with other 'classic' breeds. W. Powell-Owen (1918) adds to the claims that the Orpington was made *before* the turn of the century 'some twenty odd years ago'. (The Khaki is regarded as having emerged in 1901.) He describes it as follows:

> Runners were mated to Aylesburys, Runners to Rouens and Runners to Cayugas, and the progeny were crossed with one another until the buff plumage was secured. Blues followed by the use of those ducks which sported some blue plumage, the darkest being mated with Cayugas and Pekins.

Edward Brown (1929) was bold, or rude, enough to impute the influence of Blue Swedish in the creation of the Blue Orpington. He was also keen to dismiss the whole breed of the Orpington:

> Those who keep it are chiefly concerned with egg-production. It has made very good records in Laying Trials both individually and in flock averages. In a few instances these have been remarkable.

In retrospect, his next generalization is more surprising: 'Up to the present it has not proved popular among exhibition breeders'!

A. F. M. Stevenson (one time secretary and president of the Buff Orpington Duck Club) would undoubtedly have disagreed with him. In the invaluable little volume, *Ducks*, published by *The Feathered World* in 1926, Stevenson gives one of the most detailed examinations of the Buff Orpington. He is particularly astute in his comments on management and identification of breeding faults:

> The most prevalent faults which keep cropping up are as follows: – White feathers in neck, dark secondary wing feathers, blue lacing and brown bills in ducks, grey heads and green bills in drakes. A colour especially noticeable in ducks which must be avoided, and which is all too common, is the pale lemon buff. This colour is frequently bred from even the best parents, together with 'sports' of wild duck colour. It is a peculiarity of the breed, and seems at present to be unavoidable.

A photograph included with the text shows a 'bad specimen of a duck' with a white ring on its neck, a Runner shape, as well as lacing and barring on the wings. Stevenson was probably seeing throw-backs to some of the original Runner blood. Three-quarters of a century later there is still a tendency to get blue lacing, brown bills in the females and grey creeping on to the drakes' heads. (A reference to the genetic problems has been made in the previous section.) Stevenson goes on to say:

Fig 86 A pair of Buff Orpingtons. The duck shows the fault of a faint eye stripe.

After the first month a duck nearly always improves in wing colour, and, in fact, in general plumage. One must allow for ducks fading and showing some uneven coloured and laced feathers when they have done a heavy winter's laying, and especially after a false moult. Their bills often become brown with exposure to the sun and weather, recovering their colour after the full moult.

After a drake's early summer moult the head becomes buff or nearly so, and he does not regain his dark head until autumn. The young drakes also have not dark heads at first. This is often the case with wild birds' colouring, the male's summer plumage being similar to the female's.

Qualities

Stevenson, amongst others, regarded the Buff Orpington as the 'farmer's duck'. It was hardy, a heavy layer and a good table bird. It also had the reputation of being much more docile than the Runners, though this is largely a matter of specific strain. We have one or two exhibition White Runners that are as bold as brass and virtually 'bomb-proof', but the generalization is reasonably sound, for geese as well as ducks: 'The bigger they are, the softer they are'. Like most ducks too, including some strains of Rouen, Saxony and certainly the Runners, these are good foragers. Orpingtons should lay white eggs, the green ones having been mainly bred out early this century. An average duck, if fed properly, should ideally be able to lay 200 or more eggs. Over 300 has been possible. At nine weeks old the ducklings should weigh an average of about 5lb (2.2kg), according to Stevenson.

The Blue, Black and Chocolate Orpingtons

Blue Orpingtons are reputed to have been bigger than the Buffs: 7lb (3kg) for the drake; 6lb (2.7kg) for the duck. Few, if any, are found in Great Britain at the moment, and certainly none is being shown in the major championships. It would be very difficult to vouch for the authenticity of any modern specimen: it is almost inevitable that it would contain a mixture of Swedish, Pommern, Blue Runner or Cayuga blood.

The original Blue Orpington was being advertised before 1900. It had an even shade of blue all over its body except for the small, heart-shaped bib situated on the upper part of its breast, whilst the drake's head and upper neck were a darker shade of blue. W. H. Cook, writing in 1926, maintained that his birds bred true to colour and type, admitting that very few white, black or 'splashed' were the outcome of any one season's breeding. He goes on to comment:

> Some people confuse them with Swedish ducks, which are very much smaller, and inclined to have a good deal of lacing of a darker blue on each of their feathers.

Ironically, the modern Swedish is now expected to reach between 7 and 8lb (3–3.6kg).

In his article in *Ducks* (1926), Cook gives a detailed description (*see* box on page 78).

The Blacks followed the Blues in 1913. They were started by using black 'sports' from the Blues and later incorporating some Cayuga blood to 'set the colour'. 'They are similar to the Blue Orpington in shape, and also equal in size, their bodies being a beautiful beetle green black with a white bib in front.' Cook claims that they produced 180 eggs in twelve months and 'for their size are wonderfully active, splendid foragers . . . equal to any duck for table purposes'.

Chocolates emerged in 1918. They were a rich, even chocolate in colour and with a small heart-shaped bib. Smaller than the previous two breeds they were the best egg-layers, equal to the famous Buffs in that respect though slightly larger in size.

The Abacot Ranger

One of the genetic permutations carried by domestic ducks is a general whitening of the body plumage without its becoming completely white all over. In

Fig 87 Blue Orpington drake 'Captain Jack', the property of Mrs W. E. P. Bastard. First Dairy, 1912, etc. Unlike the Blue Swedish, Blue Orpington feathers are not laced with a darker shade. (The Feathered World, *December 1912)*

Colour in Blue Orpington Ducks

Bill – Blue. A greenish tinge permitted for the present, especially in old birds, but no trace of yellow or orange.
Head, neck, body and breast – Blue throughout, free from any bronze tint, with a white bib extending from centre of neck about 3in on to the breast, roughly oblong, about 2in at widest part, clearly defined.
Head and neck of drake – A darker blue.
The soundest coloured birds will show some lacing. For the present some white in flights to be permitted, but no white in face, and any eye streak to be severely penalized.
Eye – Black pupil with deep blue iris.
Legs and feet – Dark blue. For the present some orange or yellow allowable if mottled with blue, but all yellow or orange to be passed.

Orpington Characteristics

Colour	
Buff	Mainly buff all over, except for the seal brown head and brown rump of the drake; ochre bill and orange legs.
Blue, Black and Chocolate	No longer standardized
Shape	Slightly upright; quite meaty in proportions
Size	Light: 5–7½lb (2.2–3.4kg)
Purpose	General purpose

the 'harlequin' form, the drakes retain their beetle-green heads and both sexes have some sort of speculum amongst the secondary feathers. The pure claret bib of the drake begins to develop white lacing upon the individual feathers, and the bib itself is often divided by a central white area. Furthermore, the dark colour extends along the upper part of the drake's flanks.

'The original Campbell males had dark green head and bills, grey backs, yellow legs, pale claret breasts, black sterns and a slight ring round the neck. The ducks had greyish-brown feathers pencilled with dark brown, brown heads, yellow legs and dark bills,' writes W. Powell-Owen in 1918. He could quite easily have been commenting on the Abacot Ranger. Then, of course, there is the Silver Call Duck, the Baldwin, the Silver Appleyard, the Bantam Silver, the Whaylesbury and finally the Welsh Harlequin, which is unique within this group in having a bronze speculum. There seems little wonder that confusion should have reigned amongst breeders and judges alike, and the cause of that confusion was a mixing and re-mixing of the breeds that were used originally in that first Campbell.

When submitting a breed for standardization the main requirements are that it should 'breed true' and also be 'sufficiently different' from other breeds. The evolution of the Abacot Ranger is an object lesson in how things can go wrong and also how eventually 'virtues triumph'. By all the rules of Fate, which condemned the Coaley Fawn, the Stanbridge, the Penguin and the Baldwin to obscurity, the Abacot should have disappeared by the middle of the twentieth century. Instead it looks as if it could survive as an exhibition duck of all things into the new millennium, thanks to one or two enthusiasts in this country and in Germany. It all began before the Great War of 1914–18 had reached its conclusion.

Origin

Abacot is the name of the duck 'ranch' belonging to Mr Oscar Grey of Friday Wood near Colchester. He gave the name to a breed that showed great similarities to the famous Campbell and had a similar record as an egg-layer. In *Ducks* (1926) is found a single paragraph:

> This variety has, like the Coaley Fawn, a good record at the Laying Tests as is to be expected from a study of its origin. Mr Oscar Grey intermated some white 'sports' thrown from pure Khaki Campbells in about the year 1917. The progeny were mated to a White Indian Runner drake from a high fecund dam. At the same time Mr Grey mated other progeny of the white 'sports' to a drake of the same. The resultant progeny were light drakes of

Fig 88 This is possibly one of the first ever photos of a Streicher (Abacot Ranger) dating from the 1920s. (Photograph courtesy of Avicultura)

Fig 89 Group of young Abacot Rangers. The four ducks have a distinct brown 'hood', much prized as a distinctive feature of the breed.

> Khaki Campbell carriage and type with dark hoods, and white ducks with blue flight bars and fawn and greyish fawn hoods. Successive matings of the latter were then made and a large percentage of birds similar to their dams in type and colour were produced. The type may now be considered fairly definitively fixed, and 80 per cent of the hatches have come true. The Abacot Ranger is a good table bird, and at about ten weeks old the drake averages 4½ to 5lb in weight. As its name indicates, it is also a splendid forager and is at the same time very docile.

Keeping up the tradition of naming utility ducks along military lines (cf. the 'Buffs' in the Kent Orpingtons and 'Khaki' in the Campbells), Oscar Grey called this the 'Hooded Ranger Duck'. Rangers are commandos or military scouts presumably with foraging skills. In theory at least, any cross between a Khaki and a Runner should have all the necessary qualifications. The name did not last long; it was soon changed to 'Abacot Ranger' in spite of the clear descriptive title that could apply to the heads of both the duck and the drake.

Poultry Yearbook (1925) describes the duck in a certain amount of detail, and it is worth comparing this with what came to be the British standard by 1987. The 1925 description is as follows:

> The bill is dull dark green; the shanks and feet of the ducks are dull brown, and the drakes bright orange. The drakes have the head and upper neck dark brown; the ducks have these parts light fawn during the first year, but the fawn colour fades as they grow older, and gives place to white. In both sexes the wing-bars have a purplish black speculum, and the tail coverts are the same colour. Save for these markings the whole body is white in the duck, and white with fawn mottling in the drake.

The specification for the drake's head colouring is interesting: by 1987 an element of 'intense green iridescence' had been introduced. The drake's bill is later described as 'yellowish-green'; the female's as 'dark grey, almost black'. In 1925, however, *both* sexes seem to have the 'dull dark green' bills and, apart from the tail coverts and speculum, the female's 'whole body is white'. In fact, the early Abacot Ranger seems like merely a pale form of the Original Campbell.

Neither the 1926 nor 1930 Poultry Club Standards make reference to the breed, in spite of full standardization of the Baldwin, Penguin and Stanbridge White by this time. There seems to be little information about its popularity or why it failed to impress the Poultry Club Council. What we do know is that in the National Test of 1922–23, the only pen entered won a silver cup and laid 1,173 eggs, an average of 234.6 eggs per bird in forty-eight weeks. The Rangers beat the Buff Orpingtons, which won the silver cup in section 3, by no less than 249 eggs, and beat every pen of Runners save one. In the Wye College Test of the same season, the four birds in the pen section of the test led all breeds month after month. In March 1923, the Rangers had the distinction of putting up the highest monthly pen aggregate in each of the two tests.

As a laying duck, the Abacot Ranger seems to be the right material. Part of its success is probably the added vigour of a recent cross-breed. Nonetheless it was also a good dual-purpose bird of light colour when the existing dual-purpose breeds were dark and not able to command such a high market price. The 1925 *Yearbook* comments that the Abacot Ranger 'when plucked, is sufficiently white to pass as a small "Aylesbury" in the markets'. It goes on to praise the Rangers as 'very hardy, very active foragers, and consequently very thrifty. From the initiation of the breed they have been kept and reared in the open, and have never been housed, or given any artificial shelter of any description.'

Virtual Disappearance

The stage should have been set for another 'miracle duck' to rival or surpass the Khaki Campbell, certainly one with better carcass potential. What happened to the Abacot Ranger in Britain is something of an enigma. Perhaps it was not that different from the up-and-coming competition. As we can see in retrospect, it is very easy to create such pale-silver birds by astute cross-breeding. The original Rangers could merely have been one among many similar breeds.

Before the middle of the twentieth century, the Abacot Rangers had ceased to command much respect in Britain. Charles Roscoe, in the 1943 edition of his *Ducks and Duck Breeding*, makes the disparaging comment:

> There are a few other breeds that have been recognized by the Poultry Club, such as the Abcot [sic] Rangers, Coalvy [sic] Fawns, but they are not worth considering, being nothing more than a Khaki Campbell sport in the case of the former, and Coalby [sic] Fawns likewise, or with some Buff Orpington blood introduced, but the latter so closely resemble many pure bred Khaki Campbells, that it would take an expert to distinguish them.

It seems that it was not just politics that separated Great Britain and Germany through these dark years. German breeders took more kindly to the Rangers and, thanks to their stewardship, the breed made a vital recovery in the 1980s. Credit should go to Jonathan Thompson for spotting the clues and following up with valuable research. Similar credit should go to John Langley for having the good fortune to receive eggs from Switzerland. Sadly not everyone realized the nature of these imported birds, and this may have added to the confusion about Appleyards and Welsh Harlequins that had already been further compounded by Captain Bonnet's Whaylesbury hybrids.

John Langley exchanged eggs with a Swiss breeder – some Orpingtons for what he was told were Silver Appleyards. The imported birds, two drakes and eight ducks, turned out to be bigger than the bantam Appleyards Langley had expected. The drakes had the brilliant colouring of a male Mallard, while the females were basically white, streaked with brown and grey, 'producing an attractive silvery effect – all with beautiful deep violet wing bars'. Writing in the BWA *Waterfowl Yearbook* (1986–87), John Langley says he was ready to call them Silver Appleyards (large strain) after reading about such ducks in Appleyard's catalogue. So little information was in print about the large Appleyard, there is little wonder that Langley jumped to this conclusion. Nonetheless he built up a decent sized flock and sent birds off by train to 'remote parts of Wales or other far flung corners of the kingdom'.

It took Jonathan M. Thompson to spot the 'German connection'. According to an article in *Geflügel Börse* (1929), a certain Herr Lieker had acquired a set of Abacot Ranger Ducks in 1926. He managed to stabilize the most commonly recurring colour-pattern in 1928. (From personal experience with Appleyards and Silver Calls, we know the tremendous variation *between* strains of Silver-Mallard birds, even *within* some strains. There is little wonder then that the German version should evolve slight differences from the original British creations.) Herr Lieker decided to name the new breed after himself, Liekers Streifere (Lieker's Ranger), and in 1934 the breed was standardized in Germany under the name of 'Streicher-Ente' (the Ranger Duck). Further references by Risler (1954) and Woith (*Geflügel Börse*, 1957) confirmed the breed's origins in England in 1917 and with Oscar Grey of the Abacot Duck Ranch.

Thompson regarded this breed as being 'the most dilute variation of the Mallard colour forms'. The Germans indeed use the term '*Silber-wildfarbig*' to apply not only to the Streicher but to a strain of Runners and another of Calls. He suggested that we retain the weights as they were in 1926 (4½lb or 2kg for the female and 5½lb or 2.5kg for the male). This was despite the fact that John Langley's birds had been bred up to a maximum of 6½lb (2.9kg) to aspire to the greater size of real Silver Appleyards. A translation of the German standard was submitted to

Abacot Ranger (Streicher) Characteristics

Colour	
Drake	Green-black head, complete neck ring; body feathers mottled brown and silver; white under-colour; violet speculum *Bill* Yellowish green
Duck	Rich buff head; body white with light flecks and mottling of light and dark fawn; violet speculum *Bill* Slate
Shape	Slightly upright; medium in proportions
Size	Light: 5–6lb (2.3–2.7kg)
Purpose	Egg-layer

the Poultry Club of Great Britain and this was accepted (proposed by the BWA in 1987).

The Magpie

Pica pica is the Latin name for the Magpie, the black and white crow renowned for its obsessive collection of miscellaneous objects and universally disliked by egg producers. 'Margaret pie' is the direct route of the derivation but somewhere along the route even the 'steak and kidney pie' and the 'pie chart' are both related in meaning to the Magpie. *The Oxford Dictionary* suggests that the original 'pie' was a collection of bits and pieces, presumably within a pastry nest. So, a 'pied' object is one mottled with a variety of colours like a magpie's collection, not just black and white.

In the world of domestic ducks there has been a deliberate attempt to mimic the actual markings on the crow-like bird, first in monochrome, then in varying shades from pale blue to brown. So far, the similarity has been approximated and simplified to a neat black patch on the top of the skull and to a heart-shaped mantle on the shoulders. The main feathers are white except for the back of the body and the whole of the tail. This is a 'reasonably' stable breed characteristic but most birds in a hatch will tend to have asymmetrical, patchy or displaced markings. The perfect bird, with its cap neatly and perfectly on the top of the head, with its mantle not lop-sided nor spilling on to the secondary feathers, is a real rarity. Fortunately for breeders 'mismarks' can be spotted early, so culling or selling for pets is made easier, if that is what one is prepared to do.

When W. Powell-Owen wrote his *Duck-Keeping on Money-Making Lines* in 1918 there was no mention of the Magpie Duck. In his 1953 edition of *The Complete Poultry Book*, first published in 1924, there is a brief description and one short paragraph in which he classifies the Magpie in the 'Heavy' section (like the Orpington) and refers to it as a 'dual-purpose breed originated in Wales, bred on scientific lines for many years by the Rev. M. C. Gower Williams and developed also by Mr Oliver Drake'.

Origin

How the birds were originally produced for their striking, pied markings was never documented. An indication of how this may have been done has been given by John Hall who has observed that in large flocks of ducks of mixed origin, birds with multi-colour markings do arise from the cross-breeds. This was also the experience of breeders in Germany who created the look-alike Altrheiner Elster-Enten from crosses involving the Cayuga and Pekin.

The general consensus seems to be that black and white birds occurred for many years in west Wales before the breed was developed and standardized. The 1926 publication *Ducks* also gives credit to the pioneer work of Gower Williams who had 'bred them on scientific lines' for ten years before that. There is a picture of a 'Magpie duck and drake, both first at Olympia, and the property of that well-known breeder, Captain C. K. Greenway, Stanbridge Earls, Romsey, Hampshire'. The drake looks well marked on the mantle and tail, although the cap seems rather broad and squared off at the neck. The female, on the other hand, has a mottled cap slipping down towards the bill; there are signs of spottiness around the eyes and there is much black and white mottling behind the legs and along the stern. This would confirm the problems found by many modern breeders in getting ducks to the same standard of the drakes.

In *The Feathered World Yearbook* of 1921 Clem Watson wrote:

Fig 90 A trio of Magpie Ducks at Blackbrook Zoological Park.

> The Magpie has been produced essentially as a commercial duck, and under suitable conditions cannot fail to be profitable and a success. The Magpie succeeds wherever ducks can be kept, being at its best on open range in poor and marshy land; is a great forager, very docile, and always comes home at night. Generally comes into full lay in February, and continues with unfailing regularity until September and October. Is easy to rear – one breeder reared 285 in 1920 with only three deaths – and can safely be marketed at twelve weeks in grand condition for high class trade. Owing to its contrasting plumage it is a suitable proposition for the fancier, who can practise the art of mating to produce perfectly-marked specimens.

By 1926 a Mapie Club had been set up, the secretary being Mr J. S. Parkin, Ouse Manor Farms, Sharnbrook, Bedfordshire. Examples of the breed were submitted to the National and Wye College Laying Tests, where they distinguished themselves 'to a certain degree'. Roscoe (1941) places the birds seventh in his own league tables of egg-producing ducks, behind Khaki Campbells, Fawn-and-White Runners, White Runners, White Campbells, Orpingtons and Fawn Runners.

In *Ducks* (1926) we find praise for the table and exhibition potential of the breed. Clearly the members of the Magpie Club were not going to be as reticent as Mrs Campbell.

> The Magpie makes a very good show bird for, apart from being more than usually striking, it possesses the great advantage of improving with age, moulting out better each year. A case in point is a drake hatched in 1922, shown in the photograph (taken by A. Rice of a bird belonging to Mr J. S. Parkin) which won considerably more prizes in 1925 than it did in 1923. Incidentally this drake is considered the most perfect that has been bred yet . . .

The bird in the picture accompanying the extract is spectacularly good, and it shows none of the photographic 'touching up' that is apparent in Rice's other illustration of the Magpie in the 1930 standards.

Breeding Characteristics

By 1954 the Magpie was relegated in the standards to the 'Other Breeds' section, only to be re-instated by 1982, still merely in the Black and White form. By 1997 three colours were recognized at last: the Black and White, the Blue and White, and the Dun and White, the latter having light grey-brown markings upon the normal white ground. That it took over seventy years to standardize the blue is amazing, especially when the Magpie Club itself had witnessed the innovation by 1926 (*Ducks*, 1926):

> Two colours are accepted by the Club – Black and White, and the quite new Blue and White. The latter have been bred from sports from Blacks, and the only difference is that they are blue on back and cap, where the original ones are black. The former variety is probably the most attractive and is certainly the best known.

Before the article ends there are two little paragraphs. They outline some of the principles adopted by the early breeders.

Two matings recommended by Mr J. S. Parkin were as follows:

> (1) If you have a good drake with a nice clean cap and no black under the eye or black streaks running down the back of the neck, also with good white primary and secondary wing flights and good type (but not coarse), you can mate up to five ducks of good type and as sound wing markings as you have. But to mate with this drake I like ducks with no cap at all or very little black on heads.
> (2) The second mating is practically the reverse; if your drake is similar to No. 1, but very weak in cap, or none at all, you can mate ducks with good caps or

> heavy black in head and otherwise similar to No. 1 ducks. From either of these matings you should breed both good exhibition drakes and ducks. So far drakes come more perfect in markings than ducks, but you can breed good birds of both sexes from the same pen. *This breed does not need double-mating.*

There are certainly enough nightmares in breeding ducks without going down that particular cul-de-sac.

Quality of breeding stock, on the other hand, is always desirable. Writing at a time when duck-keeping was fundamentally a money-making exercise, Reginald Appleyard was emphatic (17 November 1933):

> Good stock costs no more to care for and keep than does inferior stock. Moreover it is a pleasure to keep and work with, and of great interest from the first day to the time you are in the proud position of breeding and owning the 1st and special winner in your local show, or even at a classic event.

He was right, with any form of stock, certainly with ducks where good ones cost little more than bad ones. The breed he chose to illustrate his argument in this edition of *Poultry Worl*d was, of course, the Magpie Duck.

He recommended to his good friend 'Mr Brown' that poultry have double benefits: those birds that are not good enough in shape, colour or markings for breeding have a sound commercial value. Profit was, after all, the main incentive for the original duck designers.

> I mention the Magpie because I consider it a most useful breed. It is sufficiently difficult to breed a decent coloured and marked specimen to arouse a 'fancier' feeling in most who commence keeping them, whilst a surplus provide good quality table meat.

Even as the ducklings dry out, mismarked ones can be identified by the quality of dark patterns in the fluff. These can then be sold as day-olds leaving the duck breeder to concentrate only on the best. Such is a great advantage to those who do not want lots of ducklings about, to those who do not want to continue feeding expensive grain and pellets to fattening rejects, and to those who want only to have obviously superior birds running around their premises.

Appleyard recommended that the breeder keep records and pedigrees – more important with Magpies than most breeds. He suggested noting which female in a group produced the best ducklings to a good drake. He then told Mr Brown to scrap the other females from the breeding-pen and toe mark the ducklings from that original successful mating. The daughters could be put back to the father to fix the good colour patterns. His main provisos for this inbreeding were that the young ducks be perfect in size, shape and stamina. He warned against the overpowering temptation to say: 'She is lovely, almost perfect; what a pity she has a slightly wry tail and is inclined to be badly dished in the bill. I'll chance it and use her.'

The wry tail and dished bill will still be there in future generations. And there is another problem with breeding such difficult birds. Appleyard notes it obliquely. With *any* bird that has very arbitrary or aesthetic requirements, i.e. colour or patterns that are decided by people rather than what is genetically probable, breeding has to sacrifice health, size and fertility to get those arbitrary patterns. Just imagine how much easier it is to breed healthy, white birds than those where only one in several hundred is anything like ideal.

Magpies are for idealists. As show birds, most exhibition specimens are unlikely to be voted 'Best Light Duck'. It is always easier to point to their obvious weaknesses, whilst plainer breeds hide behind their ordinariness; but where a Magpie is close to perfect it stands a high chance of being noticed and an equal chance of being picked as 'Champion Waterfowl'.

Continental 'Magpies'

Mix up a flock of white and coloured ducks, and an assortment of size, shape and colour will ensue, we have been told, including perhaps the odd Magpie. This certainly happened in Germany where the Altrheiner Elster-Enten (a black-and-white duck) had Cayuga and Pekin in its origin. Coloured ducks with no particular regularity of markings had existed in Germany for some time, and Paul-Erwin Oswald set out to breed a nicely marked 'piebald' duck in the late 1970s. The Altrhhe.iner Elster-Enten was developed possibly with only German breeds, or possibly with the influence of the English Magpie. Looking at Schmidt's photographs of the 'German Magpie', the only difference between ours and theirs is the rather stronger German duck's head with a straighter bill, much more like a Runner than our Magpie that has a typical utility-duck's slightly dished bill.

Other Magpie colours were also tried in Germany, according to Schmidt (1989), without much success, though German breeders complained about a brown colour being passed on from the Campbell and their difficulties in eliminating this. Perhaps the Campbell was the origin of our Dun and White Magpies, which are rather difficult to obtain now, but could perhaps be made by using the two breeds.

Conclusion

One can do little better than quote Reginald Appleyard in his 1933 'letter':

> To sum up, the Magpie is a most useful breed – a splendid layer – of good marketable size, with light bone and good quality fleshing – all white breast and under-feathering (and consequently plucking out a good colour) – a comparatively small eater with good foraging properties – eminently attractive, a perfect picture in black and white, blue and white, or dun and white.

Magpie Characteristics

Colour	White with cap, mantle, back and tail of black, blue or dun
Shape	Slightly upright (35 degrees); medium in proportions
Size	Light: 4½–7lb (2.0–3.2kg)
Purpose	General purpose egg-layer

The Silver Appleyard

Most of the designer ducks, developed by cross-breeding in the first half of last century, were for egg-laying. The Silver Appleyard is one of the few successful breeds produced for the table as well.

How and when Reginald Appleyard 'created' these beautiful ducks is shrouded in mystery. One of the only references made by the famous duck and goose breeder occurs in his promotional pamphlet issued after the Second World War. Strangely there is no mention of it in the later 1940s editions of his book *Ducks: Breeding, Rearing and Management*. The pamphlet lists Silver Appleyards among an array of breeds including 'Giant White Pekins . . . Buff Orpingtons; Blue Orpingtons; Black Orpingtons' and many more.

It is equally strange that whilst the Blue and Black Orpingtons have been consigned to obscurity, the Silver Appleyard continues to attract its devotees, especially since Tom Bartlett produced a miniaturized version in the late 1980s. Perhaps it is the name that is the attraction.

Identity

> An effort to Breed and make a beautiful Breed of Duck. Combination of beauty, size, lots of big white eggs, white skin, deep long wide breasts. Birds have already won at Bethnal Green and the London Dairy Show and Ducklings killed at 9 weeks, 6½ lbs cold and plucked.

This is from the Ixworth pamphlet, and it clearly lays out the breeding plan, to produce the best all-round utility and farmyard table duck. Appleyard certainly believed he had succeeded and there are few people who could match his expertise in the decades around the Second World War. His show medals for geese, ducks and chickens illustrate his pre-eminence.

For a good table duck it is useful to have a light-coloured carcass. The white skin of a Pekin or Aylesbury tends to look 'nicer' to some customers than the gamier looking Rouen or Cayuga. The duck may or may not dress more easily. However, unlike some of the traditional strains, a new commercial breed would have to be quick growing. The birds eat a huge amount of food, and every week of growing is an extra week of feeding. Extreme vitality and speed of growth are the keys to cost-effectiveness. There is also the tender taste of young flesh. Whilst connoisseurs may prefer the richer tastes, many consumers like the bland, but tender, meat of fast-reared ducklings. They can also tell they are not getting 'old duck'.

The Silver Appleyards, when they were first developed, had the vigour of recent hybrids. They had also a high meat to carcass ratio. The size of a goose or duck is no guarantee of meatiness. A chubby bird is a much better bet than a thin, rangy specimen. Appleyard specified that the key meat area, the breast, should be deep, long and wide, and that is still a good guide when choosing breeding stock. Too many modern specimens are inbred, with skinny heads and thin bodies. Chunkiness is essential to an Appleyard whether full size or miniature.

Then there are the eggs. Whilst many customers of hen eggs go for brown shells irrespective of flavour, a lot of duck egg buyers prefer white to green or black shell hence Appleyard's requirement of 'lots of big white eggs'. This was, after all, to be the best all-round bird on the market. But what should it look like?

Verification

Appleyard continued to develop the breed during his working life, but had not succeeded in stabilizing it or producing a standard by the time he died in 1964. The artist, Wippell, however, was commissioned to paint a pair of the ducks in 1947. These were probably from the stock of the winning birds exhibited at the Palace that year, but were painted from carcasses sent through the post. Nonetheless, the painting does give a very good representation of what the original may have looked like. It is certainly a guide for breeders, judges and exhibitors. It also gives revealing insight into the putative genetic make up of the Silver Appleyard.

Wippell's duck and drake are fairly upright in carriage and have a head shape that is chunkier than many of the specimens seen in show cages. The inference here is that a certain amount of Pekin blood has somehow found its way into the new breed. This is, of course, largely deduction and guesswork. Other than by documentary evidence, the only way to establish the viability of such breeding is to recreate the bird from scratch. It is possible that Appleyard used Rouen, or more likely something resembling a Rouen Clair, to create the prototype. Certainly the female Rouen Clair has the single pencilling on its scapular and body feathers that more closely matches the Appleyard duck.

Another piece of evidence on the original birds has recently come to light. Cyril Pilkington of Lancashire obtained two pairs from Appleyard in the 1940s and his photograph of them also confirms their similarity to the breed today. Their main

Fig 91 The BWA described this as: 'A rare historic picture of one of the first pair of Silver Appleyard ducks sold by the late Reginald Appleyard. Sent to us by Mr Cyril Pilkington of Westhoughton, Lancashire.' Guy Wilkins endorsed the photograph of the birds as the type he remembered at Ixworth. (By permission of Cyril Pilkington)

Fig 92 Wippell's painting of a pair of Silver Appleyards. The birds were painted from dead specimens sent through the post. Cyril Pilkington therefore wondered if the carriage of the birds was correct, i.e. not high enough.

dissimilarities are in the colour of the duck's bill (black in the 1940s) and the carriage, which was even more upright than in Wippell's painting. He also thought that the birds in those days were quite variable in colour and produced some white females, his comments about the carriage and colour perhaps confirming the addition of Pekin in the breed.

Several breeders have tried to recreate the Silver Appleyard and tribute must go to Tom Bartlett of Folly Farm for his successful development programmes. He pioneered the Miniature and did much to stimulate interest in the large form. He was fortunate in getting hold of what he recognized as Silver Appleyards in Chipping Norton market in the 1970s. He was also quite certain that the birds being shown at the time were not genuine Silver Appleyards (*see* The Abacot Ranger). Tom set about breeding birds to resemble the Wippell painting that he borrowed from Reginald Appleyard's daughter, Mrs Noreen Godwin. Many of the ducks now available owe their existence to the work of Tom Bartlett.

Appearance

The basic colour scheme of the Silver Appleyard is simply a lighter variant of the Mallard/Rouen wild form. Silver greys take over from gun-metal greys on central scapular feathers of the male and the lower flanks. The drake also has a distinctive silver cheek patch and faint eyebrow stripe as well as a pale throat. The claret breast markings are 'laced', unlike the Mallard form; there is a central break in the bib itself and the claret colour tends to 'run' along the upper flanks. All of this is compatible with what Lancaster calls a 'restricted mallard gene', which occupies the same locus on the chromosome as either the 'dusky' or 'Wild Mallard' genes referred to previously.

The female has also a very distinctive appearance: the head, neck, breast and body are silver white with a broad band of fawn feathers running down the neck to join the fawn of the back feathers without a break. This is a key criterion: some specimens tend to have a gap in the neck stripe and yet others have a rudimentary hood, a slight darkening of the feathers of the upper neck and head. One final weakness is the fading away of the speculum, the brightly iridescent secondary feathers of both duck and drake.

A full and detailed description of the Silver Appleyard can be found in the BWA Standards of 1999.

Problems

Appleyard's reluctance to mention his silver duck in any of the editions of *Ducks* that post-dated his 1936 pamphlet, leaves room for a lot of speculation. Did he find, for example, that the original Silver Appleyard did not breed true? This is a most unlikely scenario in the knowledge we have of its subsequent development. Also, Appleyard was far too skilled a duck breeder to broadcast his success with an unstable cross-breed.

There is one inkling of a problem, however. The stability of the breed is made vulnerable to the light/dark tendency that results in double-mating in some other breeds. After many years, during which dark females were ignored, Silver Call Ducks have been recognized in two distinct colour strains: the Silver Call and the Dark Silver Call. Silver Appleyards also tend to produce, in both the large and the miniature forms, light and dark ducklings. What happens when the breeder concentrates exclusively on one shade will obviously depend on

Fig 93 Appleyards are always busy. Five ducks splash on their pool at Tom Bartlett's in 1999. They show the cream throat and coloured head and eye marking typical of the strain. The larger feathers are pencilled with a distinct 'V'.

Fig 94 A pair of Silver Appleyards at Blackbrook Zoological Park.

the particular strain. It is likely, though exceptions abound, that the resulting progeny will polarize towards increasingly divergent forms. Perhaps one day someone will insist that the Waterfowl Standards recognize both dark and light colour types of Silver Appleyard. This would indeed be rather sad, and it would be just as unfortunate if the most desirable males and females were from the opposing strains – hence 'double-mating'.

I must confess that my idealistic explanation for Appleyard's modesty is similar to that of Mrs Campbell's reluctance to produce a very detailed standard for the 'Khaki'. The Campbell was simply a very good utility duck, and many of its supporters preferred it to reign in the realm of commercial duck-keeping. The Silver Appleyard too is designed to be a good utility bird. I can recommend it certainly as a table bird. It tastes good, whatever you call it. I wonder if that can be said about all those breeds that fill the show cages?

The Welsh Harlequin

Origins

Most, if not all, of the light ducks have got Indian Runner blood in them. This is what makes them so prolific as egg-layers and it gives many of them a slightly upright and lively posture. As we have seen, these are the 'designer ducks' that were bred about the turn of the century and many of which were given names by their originators: Campbells, Orpingtons and Abacot Rangers. They are healthy, fertile and marvellous egg producers, and so they should be: they have all the hybrid vigour of recent cross-breeds.

Few, however, have got such a chequered career as the Welsh Harlequin. For those who are keen on getting some light ducks to run around the garden or farm, and who want a regular supply of eggs, it is worth trying a flock of Harlequins. They are as good as Khaki Campbells. In fact they *are* Khaki Campbells, give or take a few genes. Just a word of warning: as with all light breeds, do not keep too many drakes. One is sometimes more than enough for a yard full of females. Two or three can be rapacious. If you want healthy ducks, keep an eye on the drakes.

As to their chequered career, it is probably best to start at the beginning. According to William Bonnet, his father gave up rearing chickens because of coccidiosis in the flock and went on to breed Khaki Campbells. Near the end of the War these were cheaper to feed than Aylesburys and they produced eggs in large numbers. These were of course in very short supply even after 1945. Group Captain Leslie Bonnet 'hatched all his own ducklings from eggs laid by his own birds. It was during one of these hatches that two honey-coloured ducklings appeared – as luck would have it, a duck and a drake.' All this was in 1949 at the family home in Flauden, a small village in Hertfordshire.

From the two birds a small flock was developed. The birds were similar in size to the Khaki Campbell, had the same egg-laying capacity and yielded a slightly paler coloured meat. At this stage they were known as Honey Campbells, a term that is worth bearing in mind in the light of future developments. The genuine Harlequins have always had a distinct honey tone to the background colour of the plumage, especially in the females.

From Hertfordshire the Bonnets moved to Criccieth in North Wales, where further developments took place. The first was a change of name around 1950. According to William Bonnet, his father showed genuine entrepreneurial spirit by calling them Welsh Harlequins to sell them to 'a wealthy local woman who kept a show farm, on which all the animals were Welsh'. A much more appealing name, this is perhaps one of the reasons for the breed's attraction. Locally named breeds have a certain bucolic resonance for the general public and the word 'harlequin' creates memories of pantomime clowns covered in bright spangles, which is not a bad parallel for the adult plumage of the drakes. The new name was a stroke of advertising genius, even if it did confuse a few people who automatically thought of the wild Icelandic duck *Histrionicus histrionicus*. The name stuck and the flock grew to as many as 5,000.

The next complication in the 'spangled' career was another entrepreneurial decision. Bonnet

Silver Appleyard Characteristics	
Colour	White ground colour
Drake	Head black-green with silver cheeks and throat; broken claret breast; dorsal surface mottled with grey and claret
Duck	Upper head and neck striped with fawn; dorsal surface fawn with streaks of grey-brown
Shape	Slightly upright; meaty proportions.
Size	Heavy: 7–9lb (3.2–4.1kg)
Purpose	Egg layer and table bird

Fig 95 A rare photo of a Welsh Harlequin duck (left) entered in the 1952 National Poultry Show by Group Captain L. Bonnet. (Photograph courtesy of Avicultura)

thought that, by injecting some Aylesbury blood into some of the flock, he could produce more of a table bird. Using a Welsh Harlequin drake and two Aylesbury-type ducks, afterwards discarding the adults then breeding persistently with the offspring, he developed another strain that he wittily called the Whaylesbury Hybrid. For many people still this is the bird they automatically think of as the Welsh Harlequin, such was its popularity. Added to the problem were numbers of Abacot Rangers that found their way to Wales. This is where the confusion tends to begin.

It is at this point one needs to look at some of the key features of the original Harlequin to distinguish it from the 'also-rans':

- the Welsh Harlequin has all the characteristics (shape, behaviour and utility) of the Campbells;
- it has a distinctive bronze/green speculum, whereas the Whaylesburys and Abacots have the typical iridescent blue of the wild Mallard;
- the bill of the female Harlequin is dark gun-metal or slate, according to the Welsh Harlequin Duck Club;
- the background colour has stronger traces of honey or russet than the Whaylesbury, which, apart from being less stable in colour, can be very pale, almost white.

As in all good dramas, one can almost anticipate the next episode. The two groups began to get less and less well-defined then, disastrously, in 1968 the flock was decimated by a fox when Group Captain Bonnet failed to shut them in securely one night. Although he tried to revive the original blood stock he finally gave in and amalgamated the pure Welsh Harlequins with the Whaylesbury Hybrid group, bringing what looked like extinction for the Honey Campbell sports of 1949. What was left he further renamed the New Welsh Harlequin!

Now for the twist in the tale: if this was a real pantomime it would need a magic wand to prevent a mini-tragedy. Enter one of Bonnet's early customers who still retained pure blood-stock from the original Welsh Harlequins:

> Unbeknown to Leslie Bonnet, a small number of the original ducks existed in the village of Rufford, near Ormskirk, Lancashire. Mr Eddie Grayson, a registered nurse and former poultry farmer, a long-time duck keeper and enthusiastic follower of Group Captain Bonnet and his methods, had purchased some selected Welsh Harlequins from him in February, 1963.

Reasoning that the Harlequins were basically Campbells with one or two mutant genes, Grayson introduced new blood into his own little flock in 1983 from the 'same source of commercial Khaki Campbells as the first'. The result was a resurrection of the breed and a saving of the day. But that is not where it ended.

The thoroughness and punctilious determination of Eddie Grayson led to the establishment of The Welsh Harlequin Duck Club, which established, under the aegis of the BWA, a rigorous standard for the Welsh Harlequin. It was submitted to the BWA Council in 1986–87, and is now accepted by the Poultry Club. It is a tribute to Mr Grayson that he was able to publicize the breed so effectively in the face of astonishing confusion and managed to gain support from all the official bodies. The story managed to have a happy ending and all the pure-breed Harlequins found in show cages or small-holdings owe their existence to Bonnet and Grayson.

As a consequence of all the publicity and hard work that went into the 1987 standard, there was an upsurge in popularity of the breed. Back in 1988, at the BWA Championship Waterfowl Exhibition in Malvern, some of the largest classes of ducks were those of the Welsh Harlequin. Competition was rife and a free circulation of stock was spurred by Grayson's willingness to distribute eggs around the country. The prime motive was to popularize and broadcast the 'true' breed as widely as possible. Many new, and some established, duck breeders took on the challenge. Gradually more and more people began to recognize the 'new' standard that supplanted the 1982 version, which stated, amongst other things, that the drake's bill should be 'gun metal coloured' and the female's 'pale yellow or khaki'. Least lamented was the assertion that the bars (specula) on the wings be 'electric blue'.

Fig 97 Welsh Harlequin Duck. (W. J. and L. C Bowen)

Appearance

The Welsh Harlequin drake has an iridescent green head with bronze lustre. Its white collar goes all the way round the neck and its breast is rich red-brown mahogany, the colour spreading along the shoulders and down the flanks. The closed wing over-feathers, the scapulars, produce a rich tortoise-shell appearance, probably giving it the Harlequin name. Its bill is green and, most important, the secondary feathers are a distinct bronze colour.

The female has a faint, honey-fawn 'hood'. Her body feathers are creamy fawn with brown central markings. Besides the bronze secondaries, the key features for the duck are gun-metal/slate bill and dark brown legs in older birds.

Fig 96 Welsh Harlequin drake – Whitchurch, Shropshire 2000. (Ray Glover)

The Saxony

One of the most attractive of the heavy ducks, the Saxony, is also one of the most effective exhibition birds. The breed has become very popular in the last twenty or so years, scoring as Champion Waterfowl in many shows – to such an extent that many people think of it as one of the great classic ducks. In fact, it

Welsh Harlequin Characteristics	
Colour	
Drake	Green-black head, complete neck ring; body feathers mottled brown; light honey under-colour; bronze speculum *Bill* Olive green
Duck	Rich honey head; body honey-cream with light flecks and mottling of light and dark fawn; bronze speculum *Bill* Slate
Shape	Slightly upright; medium in proportions
Size	Light: 4½–5½lb (2.0–2.5kg)
Purpose	Egg-layer

is one of the latest of the deliberate cross-breeds, not stabilized in its present form until the 1950s.

Die Sachsen-Ente

The Saxony duck arrived in Germany by the route of the hybrid. The breed was developed by Albert Franz in two phases. In 1924, this breeder had tried crossing the Rouen, Pekin and blue Pommern (Schmidt, 1989) to try to create a heavy, meaty duck for the table at about ten weeks old. At the same time, he wanted to breed an attractive show duck in an interesting colour. One important point for the breed was a light under-colour, essential for the appearance of dressed table ducks so that the carcass was not spoiled by dark pin feathers. Schmidt felt that, in the beginning (though this was not proved), the Buff Orpington was also used to get the rich buff colour of the Saxony.

The new breed first appeared in the Saxony County Show of 1934 in Chemnitz-Altendorf where it was great success at the show, but not widely recognized in subsequent years. Then Word War Two intervened and it was not until 1952 that Franz began to collect back together, from local stock, the line of ducks bred in the pre-war era. In 1955, four breeders showed nineteen birds at the Lipzen Show, and from that point the duck's popularity took off; they were soon at all the regional events. The breed was also used in fattening trials and one drake reached 6.0kg – way over the standard weight of 4.1kg (9lb).

According to an article in *Geflügel Bürse* (21 July 1995), entitled 'Discreet kaleidoscope of colours – the Saxony Duck' by Ruppert Lunz, Albert Franz was taken prisoner during the War and when he returned to his home at Chemnitz he had to start more or less from scratch. What stock he could get hold of from local farmers was combined with newer cross-breeds obtained using his 'tried and trusted breeding principles', the ones he had used to design the prototypes. The article does not go on to compare the two versions nor suggest that there were any noticeable differences. It merely adds that the Saxony, as we know it, was recognized in the GDR in 1957 and in the Federal Republic the following year.

Schmidt notes that because this was a relatively 'young' breed, that there were considerable variations in the breed's phenotype – its outward characteristics – due to the genetic input of the three of four initial sources of genetic material. This can be seen by comparing photographs of some of the earlier birds in Germany with the ideal specimen today. Photos in *Geflügel Bürse* show females with quite a pronounced neck ring, which is regarded as a bad fault by today's standards. Similarly, the very upright carriage of the Pekin can recur as a fault too.

The Saxony was probably introduced into Britain from several sources, in the 1970s onwards, and often showed the faults referred to. However, the drake in the 1982 standards book was a good specimen and Tom Bartlett in1990 commented that they bred very true to type.

The overall appearance of the Saxony is a strong, meaty duck. They are very rapid growers, initially overtaking the Rouens, which will eventually outgrow them and become much heavier. The Saxony drakes in particular are well filled out with feathers (a characteristic perhaps inherited from the Pekin) and look heavier than they really are.

Colour

The birds show an attractive combination of colours. The drake in full plumage has a blue-grey head and neck with a smart white collar, which must completely encircle the lower neck. The rump and undertail are a matching blue-grey but the tail feathers and curls, says Schmidt, 'should be the colour of flour'. He goes on to say that all the drakes figured in *Geflügel Börse* in 1986 showed blue curls: the original characteristic seems to have been lost. The underlying body colour is a very pale grey or oatmeal and the flanks are pale grey too. On close inspection they have the typical Mallard pattern of grey stippling or graining on a light ground seen on all the grey flank feathers of the male in Rouens, and coloured Runners and Calls. On the breast, the rusty-red feathers are finely laced with silver, but the red should be confined to the bib and not stray along the flanks (as on the male Appleyard and as it used to in many early Saxonys). The wings and coverts are grey – the speculum darker still.

Drakes have a yellow bill, which can develop a green cast with age or a black bean on the tip. These are both faults, a clear yellow being preferable. Lunz additionally warns against 'open' neck-rings, brown colour in the head plumage, lack of markings on the scapulars, rusty pigment in the speculum and a very light breast colour.

Although the drakes are the immediate eye-catchers, as a result of their unusual colour, they are quite difficult to show and do not always maintain their smart appearance for long. Brown feathers tend to come in the blue of the head and the collar can disappear when the drake approaches eclipse, becoming very 'duck-coloured', just as the wild Mallard and Rouen.

It is the ducks who maintain a more interesting colour for longer. They are a deep apricot buff shading to oatmeal on the flights. The speculum itself, however, should be like the drake's, a good blue-grey, 'dove blue' in German. On the head, the distinctive dark Mallard-pattern face stripe goes through the eye with a lighter eyebrow above, and a pale marking in front of the eye. The throat, like the Mallard, is plain and paler. Legs and webs should be yellow-orange in both sexes, but the female's beak is a browner shade than the male's.

The ducks are difficult to show too. They are 'good doers' – they need to be because they will lay up to 150 eggs per year. The females tend to get a bit fat and lose their figure and need to be kept on just the right rations for really fit condition. In with the Rouens on ad lib food they soon put on the extra weight. Lunz suggests putting them on a diet of oats and greens by the end of December. Fat ducks are less fertile. Their laying ability may be impaired and what eggs are laid are less likely to hatch.

As far as the plumage is concerned, the best colour is difficult to achieve in the female, a common fault being a 'dribble' of the throat colour down on to the breast. Lunz warns against a 'speckled over-colour', missing eye-stripes, white on the lower tail, rust on the speculum and blue shading on the back.

Breeding Hints

Ruppert Lunz (1995) says that he places great emphasis on eye-stripes in young drakes. These are visible in the first six to eight weeks but lost later in the mature plumage. The head, breast and speculum colours, he says, should be 'clear'.

Two major problems have arisen in some of the German birds: dark beans on the drakes' bills and poor quality speculum feathers. Lunz refers to the latter as being 'rough or hairy'. Both of these faults tend to appear in late autumn, during the exhibition season. The black beans seem not to appear immediately but only in the mature males after the summer moult. The 'rough or hairy' speculum feathers develop when the drakes are restricted to a confined or dark space. 'I was able to observe that animals I had shown on a Wednesday would look hairy, rough and eaten by the following Sunday.'

In spite of these reservations, Lunz recommends the Saxony as easy to look after. It can be kept almost anywhere and makes no great demands in terms of feeding or care.

Fig 98 A Saxony duck and drake.

Fig 99 Head of an exhibition Saxony female showing her characteristic pale throat and face markings, Stafford 1999. (D. and J. Keohane)

Designer Bantams

Two small breeds may be said to deserve the title 'Appleyard', but only one of them is so called in the

Saxony Characteristics

Colour

Drake	Head blue-grey with complete neck ring; breast rusty red; lower body oatmeal; upper body blue-grey
Duck	Apricot buff with prominent white eye markings and cream throat
Shape	Slightly upright when alert (0–30 degrees); meaty proportions
Size	Heavy: 7–8lb (3.2–3.6kg)
Purpose	Table bird

Waterfowl Standards. The first to be developed was by Reginald Appleyard himself and is now called the Silver Bantam Duck, previously the Silver Appleyard Bantam. The second was developed later, in a specific attempt to create a smaller replica of the large Silver Appleyard. This one is now called the Silver Appleyard Miniature.

It has been common over the last ten years for exhibitors and judges alike to become victims of confusion. When filling in entry forms for the championships, many competitors are not sure which category applies to their birds. Consequently some Miniatures have been entered in the Bantam section and a number of Bantams have been in the Miniature section. Not all judges have kept or bred the two, so there have been a few 'interesting' judgements at times.

Double Identity

To save further confusion it is useful to get a mental picture of the two birds. The Miniature is identical in all but size to the large Appleyard described previously. The Bantam has strong visual similarities to the Abacot Ranger. It is as simple as that – almost.

The 1982 Waterfowl Standards give a photograph of what was typical of the 'Silver Appleyard bantam duck' of those days. It is a beautiful little bird, longer and more slender than a Silver Call Duck, with a bill somewhere between a Mallard and a Call Duck in length. Even though the photograph is monochrome, it is easy to see where the fawn hood is clearly separated from the white of the lower neck. There is a darker 'graining' over the crown, just like the female Abacot or Streicher. The body plumage too is predominantly white except for flecks of fawn and darker pigment at the sides of the breast and more thickly on the scapulars. Also clear are the markings on the wing coverts, especially the greater (secondary) coverts where the Mallard 'bars' of black and white are transformed into dark feathers with contrasting and *even* lacing; again, just like a well-marked Abacot Ranger. Below these are the mirror feathers (the speculum), fully and clearly evident. The back feathers down to the rump are flecked and stippled with dark pigment.

This little specimen was very good indeed, but not really anything out of the normal run of Bantams in the 1980s. If they were not exactly identical, as John Hall implies in his expression 'like peas in a pod', they were infinitely less variable than the Bantams being shown in the 1990s.

The confusion returns when you look at the *written* standard: 'The bantam is a variation on the larger Appleyard, with similar colouring'. There is no detailed description of the colour markings of the Bantam, so one has to assume that there is no significant difference other than size and shape. The standard of course does not give a photograph of the large *female* Appleyard, only one of a well-marked *male* that looks nothing like a Bantam drake!

On the previous page in the standards, the colour description of the large female Appleyard is also nothing like the picture of the female Bantam; in other words, the standard was as confusing and confused as some of the exhibitors and many of the judges. What happened next was an attempt to make the birds fit the words, rather than change the writing to suit the reality. It took a skilled piece of designer-breeding by Tom Bartlett to produce a little duck that, to all intents and purposes, has 'similar colouring' to the large Appleyard – only it was not quite as small, and it was not the duck actually devised by the bird's namesake, Reginald Appleyard. It was, however, honestly called the Silver Appleyard Miniature.

Silver Appleyard Bantam

Few people alive at the end of the twentieth century have as much recall of Reginald Appleyard as John Hall. Little more than a boy when the 'Master' was already in his prime, John worked for Appleyard, and Colonel Johnson, before eventually setting up in business on his own. He has a wealth of knowledge and phenomenal recall of individual birds. John is a

Fig 100 A pair of large Silver Appleyard Crested Ducks photographed with a pair of Silver Bantams for comparison. The Bantams were photographed with the large ducks to show the similarity in colour. The females show more similarity in plumage colour than the males; the Bantam drake lacks the silver cheek and throat markings. (Photograph taken in the 1970s, courtesy of John Hall)

walking encyclopaedia of domestic water-fowl in the twentieth century, but too busy with his birds to enjoy the luxury of setting it down on paper. Below is verbatim copy of a personal letter, and one of the few scraps of evidence about the origin of the Bantam (John Hall, 1997):

> This bantam breed was formerly known as the Silver Appleyard Bantam Duck. It was originally produced in the late 1940s by Reginald Appleyard from a cross between a small egg-laying type Khaki Campbell and a white call. The resulting progeny bred true (Appleyard's description: 'like peas in a pod'), except in the late 1950s and early 1960s when a very few females were produced with yellow-gold wing bars and markings instead of the original colour. They were very attractive but have not been seen for the last ten years. The last small flock in Norfolk was probably shot in the belief that it might be a nuisance with the owners of wildfowl collections!
>
> Females would lay several quite large clutches of eggs in a season and, if required, would sit and hatch at least two clutches. They were often used to hatch and rear wildfowl, being quite tame and easy to manage. They were very popular in the 1940s, 1950s and 1960s.

Reginald Appleyard himself was not a big exhibitor of Bantam or Call Ducks, which were originally placed in the 'Ornamental' section of any waterfowl show. However, in 1950, an article from *Poultry World* (18 May 1950) includes the brief notice:

> Something new for Birmingham poultry keepers will shortly be seen in the Midlands. On view at the Warwickshire Federation of Young Farmers Clubs rally . . . will be three of the first English Bantam ducks, a breed originated by Mr Reginald Appleyard of Ixworth, Suffolk.

The duck was eventually shown at the bigger exhibitions, such as the Dairy Show at Olympia, but it took some time. Lt. Col. Johnson, writing in *Poultry World* January 1962, regretted that the domestic breeds of ornamental ducks, such as the Decoy, Black East Indian and Bantam, were rarely seen because the Carolinas and Mandarins were more colourful. However, by 1967 (Easom Smith, *Poultry World* February 1967):

> A man-made variety, the Silver Appleyard Bantam, is coming to the fore. It was created by the late Reginald Appleyard and has its admirers. Quieter in tone than the foregoing varieties [Mandarins and Carolinas] and plainer in shape because of lack of extra ornaments, it might suit those whose experience has hitherto been confined to Domestic fowl.

The Colonel also illustrated the breed in one of his articles in *Smallholder and Home Gardening* (1953). After describing the large Silver Appleyard, he commented that 'a small replica is the Silver Appleyard bantam duck, very similar in colour and markings, but one third the size'. There can be no doubt that the pictures of the Colonel's birds were the Bantams, because the photograph of the drake was taken beside the Ixworth Bantam chicken to illustrate the small size of the drake. Yet this was not a typical Bantam bird; it was actually close to the Miniature Appleyard and actually pre-empted it; a very confusing situation.

Fig 101 A Silver Appleyard Bantam drake in 1953. He looks rather long in the body, but was small (deliberately photographed beside the Ixworth Bantam for comparison). Unlike the Bantam strains before and after, he showed white face markings, which Tom Bartlett bred into the Miniature replica strain in the 1990s. The photograph in the 1953 magazine also showed a well-marked duck, rather more like Tom Bartlett's Miniature replica too. The Bantams, however, must have reverted to the plain green-headed strain (probably the way they started off). (Photograph supplied courtesy of Jack Turton)

John is quite right in his championing the Silver Appleyard. It was the original Appleyard Bantam, and it was a very popular little bird. We kept and bred the Bantams for many years in the 1980s and sometimes I regret deserting the breed in favour of the Miniature. Two birds in particular were very successful in major championships: they were affectionately called 'Reggie' (after the founder of the breed) and 'Miss App' (I don't think the awful pun needs explanation). They were even used as models for greetings cards by a professional artist and eventually retired to a good home in West Wales. Their offspring were in high demand as attractive and extremely friendly pet ducks. Even as ducklings, the Bantams had 'personality'. Neither placid nor with 'attitude', they bounced with energy. Within hours of hatching they would leap out of their brooder trays and scuttle around the dining-room. Putting them back would encourage them even more.

Fig 103 A young Silver Bantam female who went on to win Best of Breed awards in 1987. She shows the typical brown 'hood' of a young bird (rather like an Abacot Ranger). The breast feathers tended to develop more colour in the autumn. The wing coverts are well-coloured in fawn and grey, laced with cream. The speculum, like the drake's, is violet-green.

Silver Appleyard Miniature

February 1986 saw pictures of the large and miniature versions of the Silver Appleyard side by side. Volume 5 of *Fancy Fowl* published photographs of birds bred by Tom Bartlett of Folly Farm, Gloucestershire. The females were almost identical, like pieces of fine porcelain. For many breeders they were a revelation. The drakes unfortunately had not been developed to a high level of complete similarity: the Miniature still had a very wide neck ring, somewhat reminiscent of a Pied Call. The Miniature Appleyard, however, had certainly arrived. How Tom bred his new mini-duck was never fully published. When asked, he usually replied that he had 'bred down' from a standard-sized bird. By what means he did this, and using what stock, has not been completely documented.

Fig 102 Adult Silver Bantam drakes have the colour of a good Silver Call. The head is glossy dark green, and the white collar completely encircles the neck. The red-brown (mulberry) bib is finely laced with silver. The area between the shoulders has the typical triangle of stippled silver, common to all the silver-Mallard breeds, and the flanks are tinged with mulberry too.

Rising to the challenge, we too decided to create our own Miniature Appleyards. That sounds very pretentious, but it was really an unlooked for stroke of fortune. I managed to spot a motley collection of farm ducks on Long Mountain. They were white, Mallard-coloured, bibbed and Appleyard coloured; only the two Appleyard females were not much bigger than the Mallards. They were cross-breeds that my neighbour had bought from another neighbour and had bred ad lib with local wild 'drop-ins'. I drove back later and eventually arrived home minus £4 but plus one nondescript specimen, that did not much appeal to Chris's preference for classic breeds. The duck was not perturbed; she took an instant liking to Reggie, the Bantam drake, and her eventual offspring had many of the attributes of Tom's Miniatures; the females were not quite as good, yet, but the drakes were much better, with clear cheek patches and cobby little bodies. Needless to say, later generations stabilized even better, and some of them have taken 'Best of Breed' titles at several major championships.

The point is that the Miniature colour form is probably a fairly common genetic possibility related to the restricted mallard gene explained in the section on colour genetics. Bred into an existing Bantam Duck, it could be stabilized into what many people had come to expect of the small Appleyard. Yet it was Tom Bartlett who had managed to see the way first. His Miniature Appleyard was not a Bantam in the strict sense of the word. Some poultry fanciers assert that a Bantam should be as little as a fifth of the weight of the full-sized bird. The Poultry Club allows the ratio to be one to four. Against this, the Silver Appleyard is one to three (the Miniature reaching 3lb or 1.4kg, the large as much as 9lb or 4kg).

The Miniature is still very much an 'ornamental' in the same sense that the Call Duck is. Few people would set out to breed this Appleyard for eggs or for the table. The birds are much cheaper to feed than the heavy or the light breeds. They are very hardy, living outside in the fox-proof pen, foraging for worms and insects, laying in hedge bottoms and bringing off broods of ducklings without any interference (or knowledge) of the owner. They look nice, are extremely friendly and make wonderful little pets. They do not seem to suffer either from the usual problems of long inbreeding: infertility, few eggs, thin shells and deformities. Fully fed they do not fly far. One or two trimmed feathers is enough to deter most undaunted aeronauts.

It is a bonus that they *do* lay far more eggs than the exhibition Call, and they have the advantage of being as fully fleshed as a Wild Mallard, which is handy if one has an excess of poorly marked drakes. I think they taste much better too than the larger breeds, especially if you remove the breasts (plus skin) and roast them, served with a rich sauce of duck stock, port and reduced red wine, or honey and fresh orange juice.

Selecting Breeding Birds

Breeding for colour always presents problems where the breed is derived as a cross from long-established colours, for there is inherent instability in the phenotype – the outward expression of the genetic makeup of the bird. From the same parents a variety of colour intensities can be bred, which is actually quite useful as the experienced breeder can select the best mating for the breeding-pen and also the best colour for the show-pen. There are thirty points for colour in the Miniature in the 1999 BWA Standards,

Fig 104 A trio of Silver Appleyard Miniatures.

so colour is given a great deal of emphasis in this breed.

As in the large Appleyard, the Silver/Dark Silver Call and the Silver Bantam Duck, you can get light and dark variants out of the same hatch. The light variant always hatches as a yellow duck with a dark stripe (almost black) along the length if its head. In both the large and miniature Appleyards, Tom Bartlett considers that this yellow duckling with the stripe is the correct bird; it will produce the correct colour in the females rather than a dark brown. It also produces silvery males but there is also a tendency for the white throat of the drake to get too broad a band, and for both sexes to lose their blue speculum. One, therefore, has to balance this lack of colour occasionally with a darker bred strain. The ducklings look quite different from the yellow ones on hatching but when they are adult it is often the darker coloured males that do better in the show-pen because they retain more of the claret colour on the breast and scapulars. Nevertheless, if this claret becomes too heavy, then the bird also becomes a darker grey on the flanks and even the underbody (like the Dark Silver Call drake, for these have similar genes for colour). In this case, the drake is too heavily marked and should be penalized.

Because the colour is such a key point, the ducklings (unless you have a very experienced eye) cannot be sold early. The drakes will not reveal their face markings until at least sixteen weeks old and twenty weeks is better. The worst ones will end up with no face markings in some strains. Also the ducks change their colour intensity as they grow and can fool you. Some of the prettiest birds at ten weeks are too pale and uninteresting by twenty weeks. You have to keep an assortment if you want to do well at the shows with these birds.

Designer Bantams Characteristics

Colour

Silver Bantam

Drake: green-black head, complete neck ring; body feathers mottled brown and silver; white under-colour; violet speculum; yellowish green bill.
Duck: fawn head; body cream with light flecks and mottling of light and dark fawn; violet speculum. Bill yellow to grey-green with dark saddle.

Silver Appleyard Miniature

White ground colour
Drake's head black-green with silver cheeks and throat; broken claret breast; dorsal surface mottled with grey and claret.
Duck's upper head and neck striped with fawn; dorsal surface fawn with streaks of grey-brown.

Blue Ducks

When is a blue duck not a blue duck? Answer: you don't get blue ducks. You get Blue Swedish, Blue Orpingtons and Blue Forest Ducks. You get Pommerns, Gimbsheimers, Huttegems and Termondes, Blue Indian Runners, Blue Bibbed Calls, but you do not really ever get a pure blue duck. What you do get is a duck that *looks* to be blue. In fact every generation of breeders seems to rediscover the formula and out pops a 'new' breed.

We discovered one a few years ago! We had heard from a breeder of pure-breed Saxonys that some of her brood would turn out perfectly white. This had never happened to us, and we wondered if there had been some contamination of imports into her flock. Therefore, on purpose, we mated a white Pekin drake to an exhibition Saxony duck. As it happened not one of their offspring turned out to be white, as one might have been led to believe. The removal of the epistatic, recessive white genes of the Pekin produced a darkening of the offsprings' plumage. Several young drakes turned out the colour and shade of blue-fawn Call Ducks. The rest were all blue birds with bibs and varying numbers of white flights. Two females, in particular, were so well marked that they would have done better than most Blue Swedish in the waterfowl shows.

This is a common occurrence, I am sure, and similar 'creations' of blue ducks are documented over the last 150 years. Most of the so-called breeds have a number of things in common. Most have a tendency to possess a patch of white feathering on the breast. A small number of primary feathers tend to be pure white. Then there is the problem of breeding. As breeders of Andalusian poultry know very well: blue birds do not produce blue offspring. Many of the brood will be a mixture of black and 'splashed' white, a sort of patchy, silvery-coloured white. This is because what you think is blue is not blue at all and, what is more, the characteristic is not genetically stable.

Technical Information

'The self-colour blue in fowls is an optical illusion,' writes Oscar Grow in his *Modern Waterfowl Management* of 1972. It is the effect of the way we see the tiny particles of melanic and white pigments: the dark and light specks blur into an apparently uniform blue colour.

Unfortunately this colour form is highly unstable. It is produced, as Grow puts it, when the pigments 'are alternately contributed to their progeny by allelomorphic parents'. It is, in other words, a heterozygous characteristic from when two gametes with dissimilar alleles at a particular locus join in one zygote. This is basically unstable and therefore cannot produce a breed where the offspring are identical to the parents. Thus, if two blue parents are mated, only a proportion of their offspring will be blue; the others will be black or splashed white. The black and splashed, if they are mated, ought in theory to have 100 per cent blue offspring.

To maintain a blue breed it is necessary to have all three component colours. What happens in practice is that the blue form becomes recognized and the blacks and the splashed are conveniently forgotten. In practical terms this means that a large number of offspring are wasted. Like mismarked Magpies, they can be culled or sold as pets, but from a show point of view it seems a real waste of a breed, especially since the 'blacks' can be aesthetically pleasing and furthermore they have the advantage of being stable. Black parents will tend to produce black offspring. Even Oscar Grow bemoans the lack of recognition given to the Black Swedish: 'It would seem that practical considerations would, because of homogeneity, cause these black recessives to gain the wider favour, but they have never been able to acquire the popularity of the more unusual Blues'.

In F. M. Lancaster's Bulletin No. 1 'Sex-linkage and auto-sexing in waterfowl' (1977), the blue colour is attributed to two particular genes: the blue dilution gene (Bl, previously referred to as G) and the extended black gene E. E is the dominant gene, which causes solid black pigment to be laid down in

Aylesbury heads. (Jackson)

Pekin drakes.

Cayugas at Blackbrook.

Rouen Clair duck. (Bartlett)

Blue Swedish. (Bartlett)

Saxony drake and two ducks.

Streicher pair.

Two Rouen ducks.

Khaki Campbell duck. (Bowen)

Silver Appleyards.

Magpies at Blackbrook.

Crested Appleyard. (Jackson)

Coloured Balis.

White Bali. (Sadler)

Pair of White Campbells. (Bowen)

Fawn Runner duck. (Jackson)

Fawn Runner drake. (Jackson)

Fawn and white Runner.

Apricot Runner. (Burrell)

Pair of Trout Runners.

Black East Indian Duck.

Bibbed and White Hook Bills at Hasselt, Belgium.

Blue Runner. (Burrell)

Welsh Harlequin duck. (Bowen)

Apricot Call duck. (Morris)

Blue Bibbed Call duck. (Sharpe)

Mallard Call. (Soper)

Pied Calls. (Barnard)

Three Appleyard Miniature ducks.

Trio of Dutch Yellow Belly Calls. (Barnard)

Silver Call drake.

White Call duck.

Dusky female. (Stanway)

Dusky Mallard drake.

Light Apricot silver duck. (Bartlett)

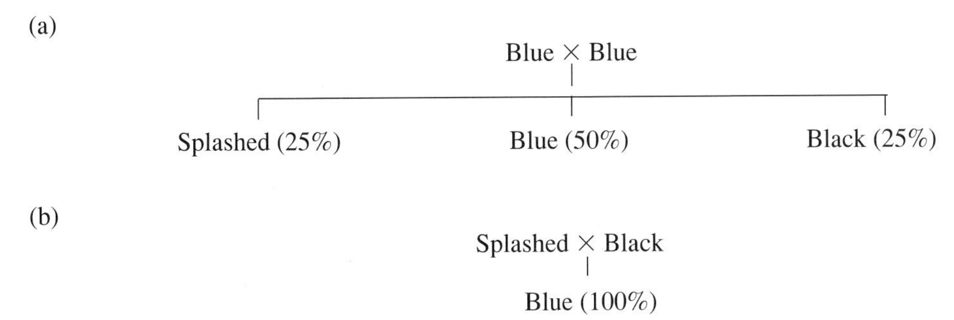

Fig 105

all areas except those for white spotting and it masks all the genes in the M and Li loci (*see* Plumage Colour in Designer Ducks). This is the gene that is responsible for the Cayuga and Black East Indian, as well as the bibbed Black Orpington. Bl, on the other hand, is an incompletely dominant gene, the effect of which is to dilute black pigment to blue-grey, but only in heterozygotes. In homozygotes the birds are much paler (splashed) and in some cases almost white.

The three possible offspring are:
Black – non-blue with extended black: (bl^+/bl^+ E/E).
Dark blue – half blue, half non-blue with extended black: (Bl/bl^+ E/E).
Splashed or pale blue – blue with extended black (Bl/Bl E/E).

So, when two 'blue' birds are mated, as in Fig 105a, the result is as shown in Fig 106. When Black (bl^+/bl^+ E/E) and Splashed (B/Bl E/E) are mated, as in Fig 105b, the offspring should be 100 per cent blue, as in Fig 107.

Historical Background

Most early references to blue ducks come from the middle of the nineteenth century at the time when the British were experimenting with Runner Ducks. In this period too the Dutch had interests in the East Indies where the tall, multi-coloured birds had their home. Whether there was any influence from the Cayuga or even the Hook Bill with dusky genes, it is difficult to surmise. The tenth edition of Bonington Moubray's *Treatise on Domestic Poultry* (1854) gives a tantalizing account of early blue breeds in Britain:

> Upon the coast part of the county of Norfolk, in the cottagers' yards are frequently to be seen specimens of a small Dutch breed, of a very pretty and peculiar appearance; the plumage is of a whole colour, either slaty-blue or dun shade, or else a sandy-yellow or cinnamon, something like the fur of a tortoise-shell cat; a gentleman of our acquaintance, who had some of the latter colour, called them Rotterdam ducks, having obtained them from that place.

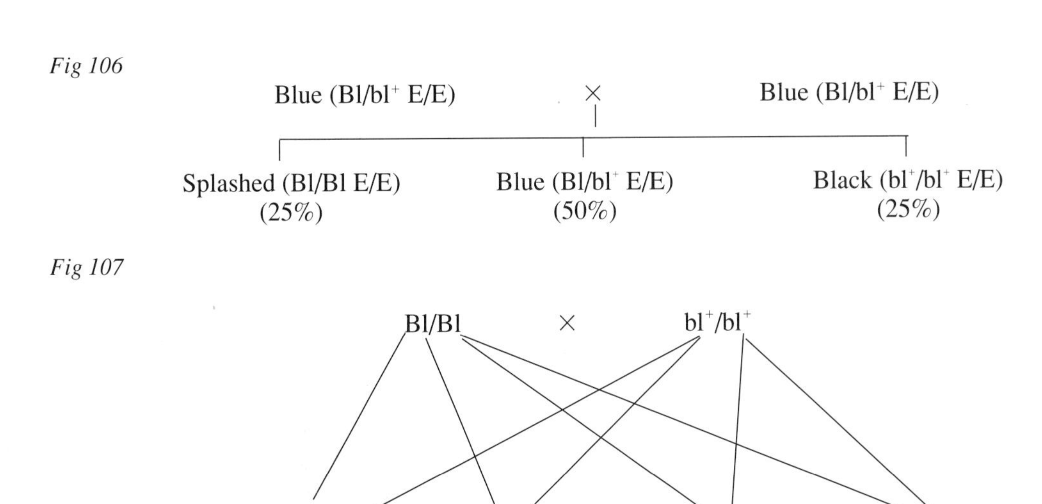

Fig 106

Fig 107

I wonder what was so 'peculiar' about the birds. Were these birds upright like Blue Runners or tiny and snub-nosed like Blue Calls? Moubray refers specifically to Calls in his next sentence, yet he makes no comparison with the Rotterdam birds. There is a remote chance that these may be related to the ancient Hook Bill Ducks. It is all very enigmatic.

Rev. E. S. Dixon writing in his *Ornamental and Domestic Poultry* (1848) also notes slate-blue and bluish-dun birds in Norfolk at this time, whilst Lewis Wright (1902) says: 'The late Mr Teebay several times told us that, about 1860, there was a recognized local race of large blue ducks in Lancashire.'

In spite of these early references, most credit for the development of blue ducks goes to the coastal regions of Germany, Holland and Belgium. Pomerania, in the area of North Germany between Denmark and Poland, is the supposed birthplace of both the Pommern and the Blue Swedish. For many years much of this area was under the power of the Swedish throne. The Termonde comes from the valley of the Scheldt between Ghent and Malines, whilst the Huttegem, according to Edward Brown, was bred extensively for many years in the Audenarde district of West Flanders.

Early references to any of these breeds are very limited but many agree in certain fundamentals. 'Blue, or slate-coloured ducks, have been common enough in different localities in this and other countries, and in each district they have acquired a local name,' wrote J. W. Hurst in *Utility Ducks and Geese* (1919).

Others speculate on the origins. *Ducks* (1926) is bold enough to assert:

> The Blue Swedish is a variety standardized in America, but specimens similar in appearance are probably not unknown in other countries as the result of a cross between the Rouen – which it closely resembles in build – and the Pekin or Aylesbury. It is not unlike the Blue Orpington, and like all blue varieties, there is probably the usual uncertainty in breeding.

The only way to support this hypothesis is to try to re-create the breed by the same means, yet it is not beyond possibility that most of the blue breeds have involved considerable experimentation, importation and cross-breeding.

Poultry Yearbook (1925) goes further, stating that with the Blue Swedish 'there is little doubt that it was employed in the production of the Blue Orpington, since the colour is very similar – blue body with a white bib'. The same article asserts that cross-breeding was involved too in the development of the Huttegem: 'This is a popular breed in Holland, and appears to be a cross in which Indian Runner blood is prominent . . . One of the ancestors of this breed is said to be the Blue Termonde, a Belgian breed'.

According to Edward Brown, with the Huttegem it is 'assumed that this race is the result of crossing the Blue Termonde, or a progenitor of that breed, upon a smaller duck with a long and narrow body of the Runner type'.

Other complicating factors, at the beginning of the century, were the lack of specificity in terminology and the extreme willingness to 'improve' the breeds with a little cross-breeding. A clear example of these can be seen in the 1910 *Yearbook* of *The Feathered World*. W. Bygot refers to Buff and Blue ducks as though they are specific breeds, which is confusing when the Orpingtons were being developed at this time. He writes: 'The Blue and the Buff Ducks are rapidly gaining favour, and we have now a specialist club for breeders of these two varieties'. After describing the exhibition performance of the Buffs at the 1909 Dairy Show, he discusses the Blues and goes on to state:

> These varieties originally came from Holland, but have been greatly improved since being imported. For utility, they make a capital cross with the Cayuga. There is a great demand for the Blue duck this year, in fact, the few breeders have been sold out some little time ago.

What is patently clear is that many of the blue breeds are related; there is often little to distinguish one from another; cross-breeding with Runners and

*Fig 108 Huttegem Duck. (*The Feathered World, *1907)*

Cayugas, in particular, is a likely component in their history and an ad lib exchange of blood-lines has gone on at many times in the past. It is almost impossible to follow an undisputed pedigree in any of the British blue breeds and nothing seems to be known about the *first occurrence* of the blue duck. Bearing these uncertainties in mind, it is worth looking at one of the latest attempts to create an all-blue duck.

Gimbsheimer

Georg Richard Oswald is credited with the origination of this blue breed. According to Horst Schmidt (1989), Oswald inadvertently crossed one of his blue females with a male Orpington. From two hatches in 1964 he obtained four blue birds: two ducks and two drakes. These he exhibited and called 'Andalusian Ducks', presumably after the blue chickens of the same name.

He had originally wanted to create a blue American Pekin, but found difficulty in getting the right bill and leg colour. What he ended up with, however, was not short of controversy. At one stage it was suggested that the new breed be called 'Blue Campbell', but since there was no direct connection with the Campbell at that time, the name was fortunately dropped. Then there was a problem with the Andalusian title. Several breeders wanted to see the import papers for these ducks from Spain. Of course there were none, so a compromise was reached by calling the bird after Oswald's home town of Gimbsheim on Rhine, which lies between Mainz and Worms. 'Wonnegauer Duck' was also considered, but 'Gimbsheimer' it remained.

One of the early imponderables was the exact nature of the blue female that mated with the Orpington drake. Apparently Herr Oswald had been attempting the same mixture of American Pekin and Saxony that we tried innocently in 1998. The number of fit, plump, blue birds from this crossing is certainly very high. It would be interesting to go through the process again but with a much greater quantity of offspring to increase the gene pool. One defect though that comes with the crossing is a tendency to produce brownish chest feathers. Another is the dominance of white bib and pinions. This problem is referred to by Oswald himself.

The Gimbsheimer is described as being of typical 'land duck' in shape. Emphasis is placed on length and breadth of body, but without keel or distended under-parts. It should not be as upright as a Campbell nor as compact as the Saxony. It should have tight plumage, a moderately arched neck and the horizontal tail feathers should be as close as possible without appearing pointed. The legs are dark red to black in colour and the bill should be grass green in the drake, black green in the duck. Each should have a black bean on the end. Feather plumage was hoped in the beginning to be a uniform blue grey ('dove blue') but in 1983 Oswald allowed a light lacing on the feathers, claiming this stabilized the colour and prevented more black feathers. Both sexes show a darker blue in the neck region, almost charcoal in the case of the drake, which is over-all a shade of blue darker than the female.

Because of the limited gene pool, experiments have taken place crossing these birds with Campbells, but the brown feather colour is so recurrent such combinations are best avoided. Also a final problem may concern the basic colouring. Some blue and buff birds are very vulnerable to sunlight oxidation. The pigment is easily bleached out and the feathers can further become stained by the swimming water.

Designer Blue Genes

The above is an exemplary tale of going round and round in circles, something which must have happened many times in the last 150 years. The Gimbsheimer was developed from the Saxony, and the Saxony was developed from the Blue Pommern. (Starting in 1924 and completing the job after the Second World War, Albert Franz eventually stabilized the colouring of a Rouen–Pekin–Pommern cross to produce his famous Saxony Duck.) Georg Oswald succeeded largely in isolating the very material that Franz had taken so much trouble to blend. One is tempted into wondering whether the Blue Orpington had gone by a similar route; also the Huttegem, which was similarly reputed to exist in both 'blue and tawny' colours.

In Great Britain, apart from the buff-coloured Saxony, only one large variety of blue duck is now accepted in the BWA and Poultry Club Standards. This is the Blue Swedish, recognizable for its neat, white bib and two white primaries on each wing, but how near this breed has ever been to Sweden, and how pure it is from Pommern, Orpington or Termonde blood, must be left to speculation. There are still Blue Orpingtons turning up. These were standardized in the 1920s and 1930s, but are no longer recognized. They may even yet be playing a role in the spontaneous generation of blue ducks.

As with Levi's and Wrangler jeans, there is much to be gained from capitalizing on other people's discoveries. Most designer ducks, from Khaki Campbells to Buff Orpingtons, have been the result

Fig 109 Blue Swedish drake.

of lucky mixtures of genes, usually beginning with the Indian Runner. In 1914 Harold Andreae set about producing a bird that resembled a cross between a Black Swedish and an Indian Runner. Black, apart from throat, breast and outer primaries, the Penguin Duck was actually standardized in 1926.

Part 2

Adult Ducks: Behaviour and Management

3 The Origin, Physiology and Behaviour of the Mallard

Introduction

Birds are believed to have evolved from reptiles, and they probably branched off from that group about 150 million years ago. *Archaeopteryx* is an early bird-like fossil, also an off-shoot from the dinosaurs, dating back to late Jurassic times. It shows a mixture of reptilian and avian features in its skeleton, as well as having the development of feathers. The leg structure and toes are similar to birds. The wing feathers, because of their vane asymmetry, do suggest that their purpose was for flight. There are adequate attachments on the bone structure for flight muscles and a small bone has recently been identified as the missing sternum or breastbone. It is short and narrow but shows some flanges for muscle attachment. Feathers themselves probably developed as specialized scales, perhaps initially providing the advantage of insulation. They have ultimately allowed the birds not only to develop flight but also to insulate themselves and regulate their temperature, and thus become warm-blooded.

During the Cretaceous, three evolutionary lines of birds evolved from the possible Triassic ancestor. Two of the these were to become extinct, but the third more recent group, the *Neornithes*, was to branch out into the modern orders of birds in the mid-Cretaceous. The ratites, gamebirds and waterfowl evolved fairly early in comparison with other groups such as parrots, pigeons and grebes. Todd (1979) considers that the origin of waterfowl dates back perhaps 80 million years, the earliest recognizable remains of wildfowl being 40–50 million years old.

Classification of Waterfowl

The animal kingdom is divided into classes, one of which is *Aves* (the birds). In this class, there are thirty-one orders of birds. The order *Anseriformes* contains the family *Anatida*e, which includes all the swans, geese and ducks. The group of dabbling ducks, to which the Mallard belongs, is the largest tribe of the subfamily *Anatina*e. All domestic ducks are related to the Mallard or its subspecies, so the physiology and habits of the domestic duck are closely related to its wild relative.

The dabbling ducks form the largest group of the waterfowl tribes and the genus *Anas* consists of 38 species and 52 subspecies. This group includes familiar duck names such as Teal, Pintail, Widgeon and Shoveller, as well as the ubiquitous Mallard. Males are referred to as drakes and females as ducks. The wild Mallard is distributed throughout most of

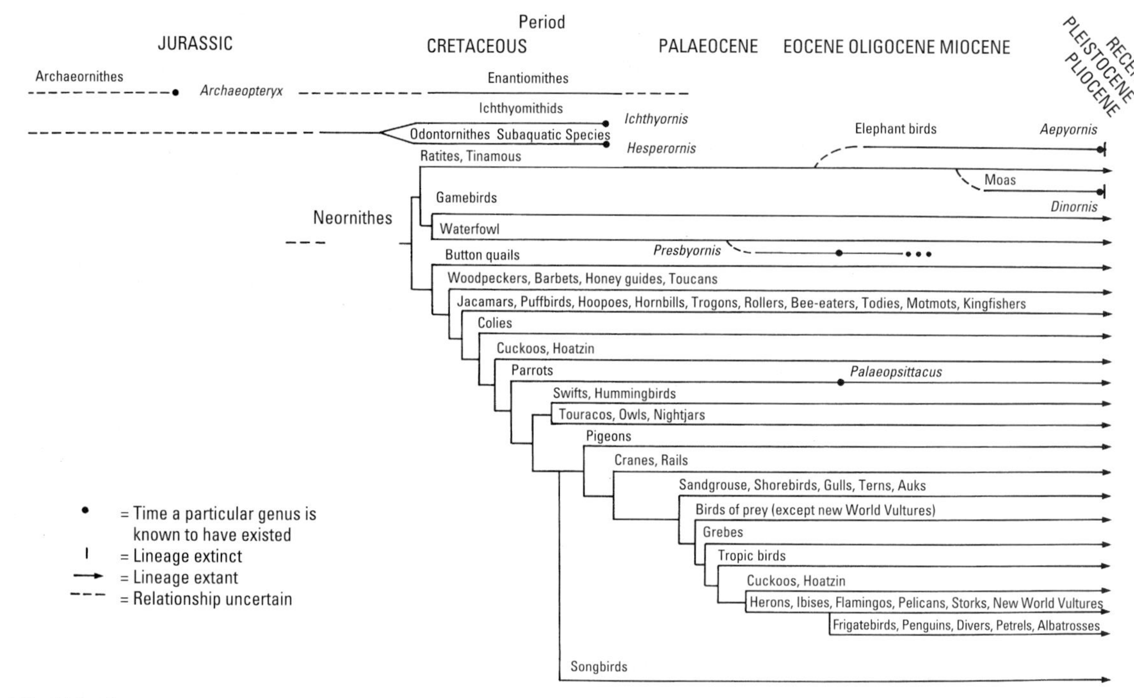

Fig 110 The approximate time of evolution and the relationship between the principal bird groups as suggested by recent DNA hybridization studies. (Dr Mary C. McKitrick in Brook and Birkhead, 1991)

Classification of the Mallard

Class – *Aves*
Order – *Anseriformes* (thirty-one orders of birds)
Family – *Anatidae*: swans, ducks and geese
Sub-family – *Anatinae*
Tribe – *Anatini* (five genera and forty species, including the dabbling ducks)
Genus – *Anas*
Species – *platyrhynchos*

the northern hemisphere. It has also been introduced to New Zealand and Australia. It has several closely related species or subspecies such as the Laysan Teal, the Greenland Mallard, Mexican Duck and North American Black Duck.

Physiology and Behaviour of the Mallard

Swimming and Diving

All the dabbling ducks have webbed feet to aid their propulsion on and in water. Dabblers, and therefore the Mallard, have their legs situated further back in their bodies than birds adapted to land life. As well as finding their food on land, Mallards spend much of their time on water, consequently the legs are placed mid-way in the body for a central position in swimming and paddling in shallow water. The set of the legs on the body means that ducks tend to waddle when they walk on land as the centre of gravity is shifted to maintain their balance.

To aid their search for food, Mallard will 'up-end' in 8–12in (20–30cm) of water to find food on the

*Fig 111 Mexican Duck (*Anas platyrhynchos diazi*) at Blackbrook, Staffordshire. Range: Highlands of Central Mexico and the upper Rio Grande Valley from El Paso (Texas) to Albuquerque (Mexico). These ducks illustrate the extensive range of the Mallard and its subspecies.*

Fig 112 Pintail (left), Mallard and Teal all up-end in water and can feed at different depths according to the length of their neck. Widgeon (right) graze or feed in shallow water. (Owen, 1977; illustration by Joe Blossom)

bottom. Paddling with the feet allows them to maintain a vertical position whilst feeding in this way. This is in contrast to the diving ducks, where the legs are set even further back to aid propulsion under water while they search for food at depths of a metre or more and stay completely submerged for half a minute.

Insulation: Feathers, Down and Fat

Waterfowl take advantage of the watery environment for feeding and protection. To cope with this medium, they have certain adaptations that must be borne in mind when considering the environment in which to keep domestic ducks.

Close and frequent contact with water means that ducks need to be kept warm, especially in northern waters where it is fairly cool throughout the year and there is also the additional hazard of freezing in winter. Insulation is achieved in two ways. First, by having two layers of feathers, but also by developing an extra layer of fat. The outer feathers are supple and glossy, maintained by preening to form a waterproof layer so that water runs off as easily as 'off a duck's back'. Below these, a soft layer of down insulates the duck from the cool water. The downy feathers have herl, which, unlike the top feathers, does not interlock. The birds cannot afford to lose condition. Un-oiled, dirty feathers become waterlogged so that a bird gets chilled and can even drown. This is a serious problem in a few domestics, which suffer from 'wet-feather'.

Fig 113 Body feathers are searched for lice and the longer, stronger feathers are drawn sharply between the serrations of the bill to comb them into place (White Call Duck).

Prior to preening, a duck will have a wash in water, going through a series of ritual activities. The head is ducked in the pool and water thrown over its back. Wing-beating, diving and flying round in circles may take place, the whole activity usually ending with a wing flap and grooming of the flights. Then the duck will stand on dry land to flick and shake the water off the external plumage prior to preening its feathers.

The outer feathers are covered with a layer of duck oil, which is obtained from the preen gland. This is situated on the rump and consists of a gland surrounded by a tuft of oily feathers. The oil is transferred by the bird first rubbing its head over the gland, and then rubbing its head over its feathers. As well as the stroking and rolling action of the head, the bill is also employed in cleaning the feathers and combing them into place. The barbules of the feathers are combed so that they interlock to form a barrier. Feathers that are apparently damaged can be reconditioned by ducks that have a good diet and clean water.

Preening therefore refers to a complex activity of combing and spreading oil over the feathers for waterproofing. It also includes maintaining the new feathers as they grow by removing the protective sheaths as they sprout from the 'pins' or blood-quills. Ducks need to spend a considerable time every day on these activities to make sure that they are weather-proof and ready for flight. During the preening process, ducks may also have a scratch, particularly round the head region. Sprouting feathers sometimes irritate, and parasites such mites can congregate here. Body feathers are also searched and feather lice are removed by passing the feathers through the beak serrations, which act as a comb. The final stage of the preening process ends with a fluffing up of the feathers, which may fulfil the dual function of insulating the feathers and down with air again, and also settling the feathers into the correct position.

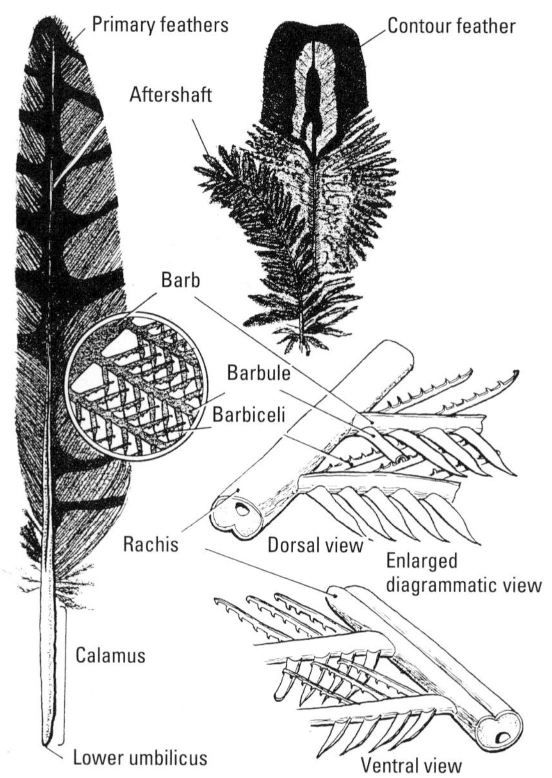

Fig 114 The structure of the outer contour feathers and a flight feather (primary feather). The rachis carries the vanes of the feather. Each vane is composed of a row of barbs linked together by smaller branches called barbules. Even-smaller hooks called barbiceli zip the feathers together. (From Freethy, 1982)

In addition to their feathers, ducks also have a thick layer of subcutaneous fat. This layer is more apparent in older birds than in young ones, where more of their energy intake is invested in growing rather than in laying down protection at that point in their stage of growth.

Ducks' feet appear to be especially vulnerable in freezing weather and exotic species from warmer climates must be looked after carefully in these conditions. The Mallard and the domestics are, however, quite well-adapted to cold. Whilst resting on land or water, one leg and foot can be hooked up into the flank feathers where it is enclosed in a warm pocket of down. The birds characteristically manoeuvre their foot into position by a series of rapid scratching movements before finally hitching it up. They are good at balancing on the one leg but in severe conditions the birds simply sit down on the ground and allow their well-insulated breast to protect them, whilst they hook both feet into their feathers.

The circulation of the legs and feet themselves is adapted to cope with the cold. Loss of heat is minimized by a heat-exchange mechanism. Arterial blood flowing to the feet is already cooled; the warmth has been exchanged to blood returning to the core of the body so that too many calories are not lost.

Moulting

Feathers have a limited life, despite their care, so they are replaced in stages at particular times of the year. As new feathers form, the old feathers fall out or are pushed out. The emerging feather sprouts from a quill filled with blood, and the blood is gradually withdrawn and sealed off as the feather achieves its mature shape and hardens.

Flight feathers are essential for escape, so these are only replaced once, after the duck has sat and incubated her eggs. She and the drake are both flightless at this time because all of the primary and secondary feathers are lost more or less at once, at the same time as the ducklings are on the ground.

The body feathers are replaced twice a year and tend to be moulted in stages. This evens out the stress on the bird's food reserves as it grows the new plumage. The autumn moult results in the nuptial plumage of the drake, the winter and spring colours of the wild Mallard, where he has the characteristic green head and white collar. In summer, however, these feathers are replaced by what are colloquially known as 'duck feathers' in the drake's eclipse phase. The smart Mallard drake assumes the sombre plumage of the female and he looks remarkably like her. In the wild this is a sound strategy; this double moult ensures that feathers stay in good condition and the colour change for the drake into female camouflage pattern gives him a better chance of surviving predation at the flightless stage.

Feeding and the Digestive System

Wild ducks are generally more active in the early morning and evening, preferring to rest during the day. Domestic ducks similarly tend to have an active period as evening approaches, and birds left out in a fox-proof pen are certainly very active in foraging on clear, moon-lit nights.

The Mallard has a broad beak, called a bill, which is a distinctive feature of any duck. The bill is quite long and spatulate, and is designed for shovelling up material. It does, however, have a harder bean at the end (perhaps for protection in shovelling) and it is covered in a soft membrane, which is replaced with wear. The sides of the bill have sieves or lamellae (serrations) along the margins. These are designed to catch particles of food such as small aquatic life and

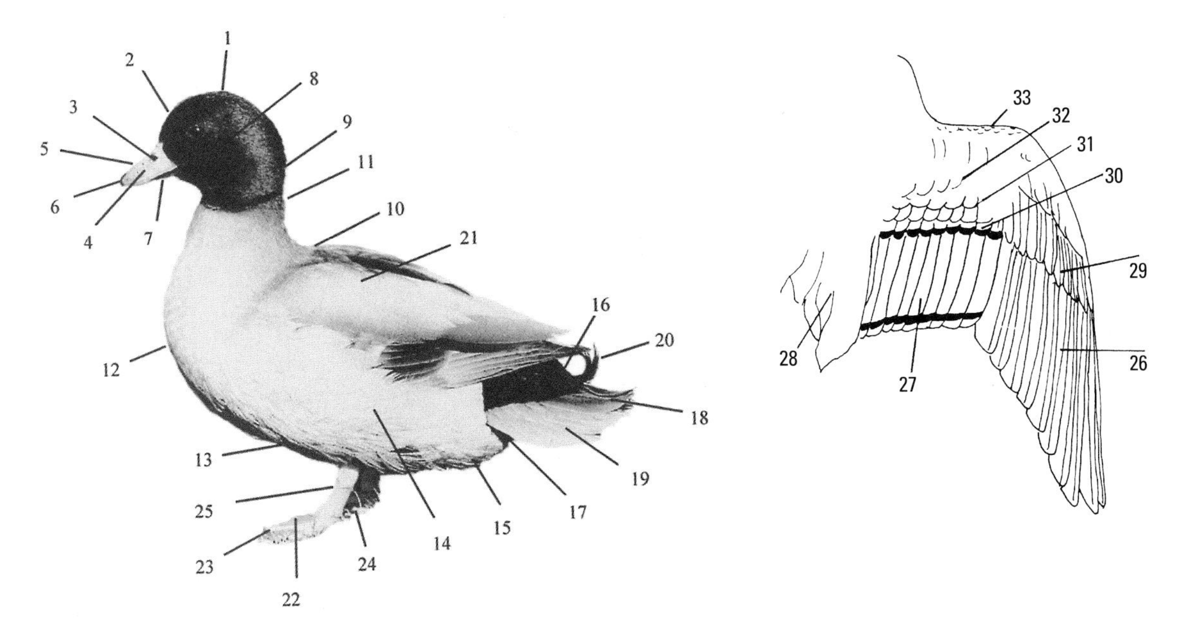

1 crown
2 forehead
3 nostril
4 upper mandible
5 culmen
6 bean or nail
7 lower mandible
8 ear
9 nape
10 shoulders
11 neck
12 breast
13 underbody
14 flanks
15 stern
16 rump
17 undertail
18 upper tail coverts
19 tail feathers
20 sex feathers (in drake)
21 scapulars
22 web
23 toes
24 hind toe
25 shank

Wing feathers:
26 primary feathers
27 secondary feathers
28 tertials
29 primary coverts
30 greater coverts
31 median coverts
32 lesser coverts
33 marginal coverts

Fig 115 Physiology of the domestic duck and the Mallard.

seeds as water is sucked in through the front of the bill, passed over the tongue and out through the sieves to strain out the food. The tongue itself is very sensitive and is lined with horny projections to catch food. Ducks will deliberately paddle their webs up and down in shallow water and disturb invertebrate bottom-dwellers to make them easier to find; domestic ducks such as Calls will mimic this behaviour.

Ducks catch and consume animal protein in the form of insect larvae and insects, snails, slugs, worms, crustaceans, tadpoles and even small frogs. The Mallard in particular is a non-selective feeder. Ducklings snap up emerging midges and mosquitoes, particularly in the evening. They also eat large quantities of vegetable matter, and even well-fed domestics will pull and keep down grass in a garden to a certain extent.

Unlike the chicken, the duck does not have a crop to store food, but the gullet (oesophagus) will do this. Domestic ducks that are accustomed to being fed at particular times of the day will stuff themselves so that their chest stands out when the gullet is full. This food is then passed down to the proventriculus, where glands supply gastric juices to start to break down the food. Muscular contractions then pass the food and the juices (a mixture of hydrochloric acid and pepsin) into the gizzard.

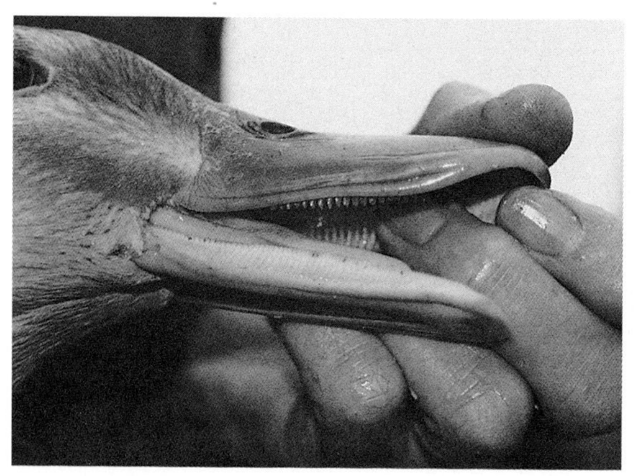

Fig 116 Head of a domestic duck to show the sieves or lamellae along the margin of the bill. Dabbling ducks such as the Mallard have more lamellae than diving ducks. These sieves, and the rough tongue, strain out food from the water or mud.

Like geese, ducks have to grind down a lot of vegetable matter so that the food can be assimilated by the gut. The gizzard itself is a large muscle where two surfaces, together with sand and grit, mill up the food before it passes through to the loops of the intestine. Here, the nutrients are digested by enzymes and secretions from the liver and pancreas then absorbed by the lining of the intestine walls. As in the goose, there are also two sacs at the end of the intestine known as the caeca. It was once thought that these were involved in the breakdown of cellulose with the aid of the bacteria found there. However, it is now generally acknowledged that birds cannot store large quantities of food, which would affect their weight and flight, and that they therefore cannot digest large quantities of cellulose like the cow (Mattocks, 1971).

Social and Sexual Behaviour

Wild Mallards flock in the autumn and pair up when the males show their breeding plumage, but this is for one breeding season only (Owen, 1977). The birds feed, rest and swim together but there are often problems because drakes normally outnumber ducks, and there may be considerable competition for females. Females may be pursued by drakes, particularly in the breeding season, and especially when they are broody and out feeding. The unpleasant side of the Mallard is that a group of males who are surplus, or whose ducks are sitting, will pursue and rape single females. They may cause such damage that the female will die or be drowned. Domestic Call Ducks in particular behave like Mallards in autumn, probably because it is unusual for Calls to lay before spring. If they are well paired and every drake has a mate, they can be kept amicably together. Once the Calls come into lay, they can be as badly behaved as the Mallard and, although certain groups will live peacefully together, the drakes can be rapacious too.

In the breeding season, Call drakes more than other domestic breeds mimic the display behaviour of the wild Mallard. They will dip their bill in the water as the head is shaken, then pull their head in sharply to bend their neck – the characteristic grunt-whistle courtship display of dabbling ducks described by Johnsgard (1992). Drakes also make characteristic head-up-tail-up movements accompanied by whistling; males may do this together as they get in fettle for the breeding season.

Wild females also indicate their interest in breeding to their partners by an inciting display which, in the Mallard, involves the female turning her head to the side several times as if she were sweeping something backwards with her bill. Again, it is female Calls (probably because they are well

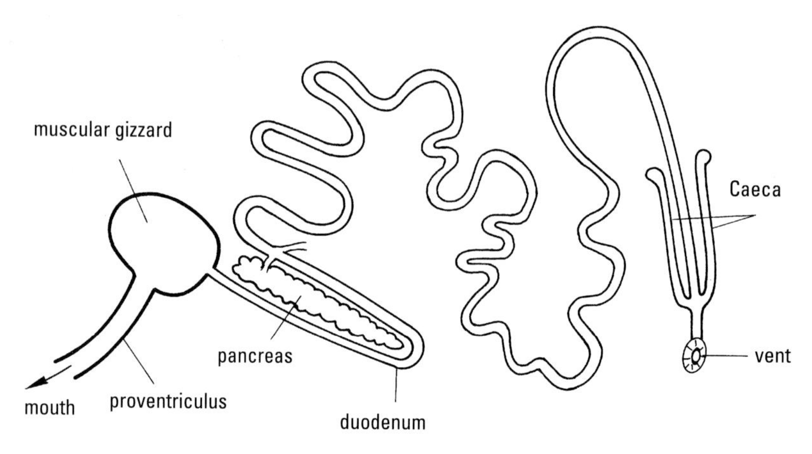

Fig 117 Digestive system of the duck. Food is shovelled up by the bill and swallowed by a series of jerking movements (not unlike a lizard's eating habits). Food passes down the oesophagus to the gizzard. The proventriculus is a broadening of the oesophagus, before the gizzard. Grit in the muscular gizzard grinds down the food so that it can be assimilated in the gut, aided by secretions from the pancreas and liver.

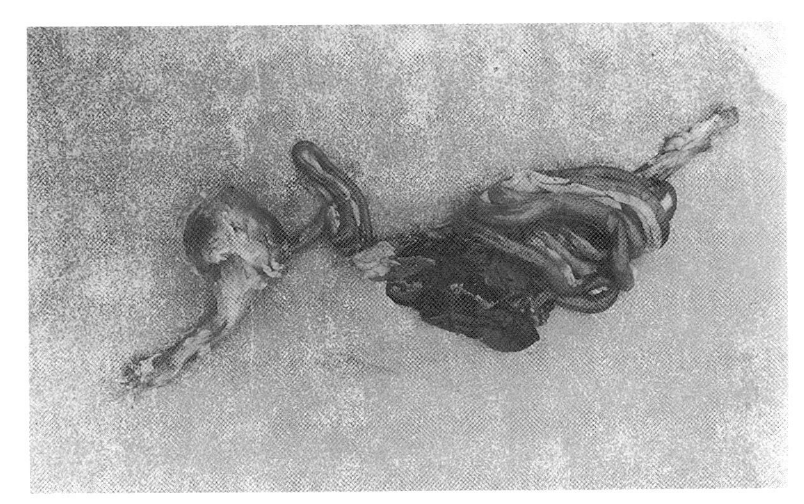

Fig 118 Photograph of the digestive system from the proventriculus (left) to the vent (right).

Fig 119 ***1****. One of the most characteristic courtship displays of dabbling ducks like the Pintail (illustrated) and the Mallard is the grunt-whistle, where the male arches his body and neck forward, flicking up water droplets with his bill.* ***2****. This is followed by the head-up tail-up after which the head is turned to the courted female. Finally the female incites her potential mate. (Owen, 1977; illustration by Joe Blossom)*

paired) that are particularly overt in this behaviour towards their drake on a pond. In addition to this display, domestic females run alongside and bob their head up and down to their drake whilst repeatedly quacking. This behaviour in breeds such as Indian Runners can embarrass a single drake beset by several females, or result in a female being set upon by several drakes if they have been left with too few ducks.

Ducks, geese and swans have highly ritualized courtship behaviour, perhaps because the co-operation of the female is essential for success of a single male for two reasons. In all birds, copulation is difficult as the male has to stand on the female's back while she moves her tail out of the way. In waterfowl, copulation is rather inelegantly known as 'treading', as the male literally has to stand and manoeuvre his position on the duck's back. This frequently takes place on water because this is their habitat. This unstable medium may make copulation more difficult, yet in the case of some of the heavier domestics, actually makes copulation more successful as the water takes their weight.

Second, in most birds cloacal contact is sufficient for fertilization but in ducks, geese and swans, a coiled penis directs the semen into the female. This perhaps ensures that the sperm is not lost on the water and may be a reason why the waterfowl have evolved in this way.

Fig 120 In the breeding season, females run alongside their drake whilst bobbing their head up and down, and repeatedly quacking.

Nesting

Wild duck nest on the ground, so their camouflage has to be good. The nest is merely a scrape in the ground made by the bird turning around and scratching at the earth with her feet. Grass and leaves may be pulled into the nest to hide the clutch of up to twelve eggs. Broodiness is the signal to add down to the nest. This insulating material is added from the duck's underparts and breast to hide the eggs and keep them warm when she is off the nest to feed. Species such as the Eider do not eat at all whilst sitting, but domestic ducks should leave once a day to eat and drink.

The incubation time of the Mallard is the same as the domestics – twenty-eight days, except for the Call which is less at twenty-six to twenty-seven days. Wild drakes take no part in their upbringing – they are even a nuisance to broody ducks, when they go out to feed – and domestic drakes too are generally best kept apart from ducklings.

4 Purchasing Stock

Why Keep Ducks?

Feare (in Brooke and Birkhead, 1991) points out that the Greylag Goose was domesticated over 3,000 years ago in the Oriental region centred on India. Later, Mallards were probably domesticated in Mesopotamia. Domestication of these birds probably had to wait for people to be settled, but also required that the indigenous species be amenable to domestication. Both the duck and the goose fulfilled this requirement; they produce fluffy precocial young, which are attractive to look at and are more likely to survive with foster human parents than naked hatchlings. In addition, both species readily imprint on real, or surrogate, parents. They could therefore easily be tamed.

Mallard are very widespread. Domestication of this species undoubtedly took place in many different parts of the globe. Centuries of selection in the Far East, North America and Europe resulted in the production of a particular type of duck favoured in each area. The Chinese valued the quick-maturing, white Pekin; in Indonesia the egg-laying capacity of the Runner was more important that the flesh. Yet in the Far East ducks were also probably valued for their ornamental qualities. Not only did the Chinese and Japanese frequently depict the Mandarin in their art but the Call Duck too may have originated from this area.

In Western culture the duck has functioned both as a utility egg-layer and table bird, but only in comparatively recent times. Although there is reference to ducks having been kept as ornamentals as a 'paddlinge' in perhaps medieval times, there seems to have been little selection before the 1800s in the USA, and the 1700s in Britain. In both cases the table duck was the main aim. The egg-layer later became a large-scale commercial reality after the Indian Runner's gene pool was used to produce several types of the twentieth century 'designer ducks'.

Today, ducks do not seem to be as widely kept on a commercial scale as they were in the hey-day of the Aylesbury trade, or in the 'designer' craze of the early twentieth century. They are reared in large numbers by a few commercial enterprises, but the battery hen and the broiler chicken provide most poultry products for the mass-market.

Domestic Ducks are Easy to Keep

- Domestic ducks are easier to keep than wildfowl, where keepers may be dealing with different nutrition and behavioural patterns with several species. Wildfowl also generally need a fox-proof pen, whereas domestics can be housed at night.
- Ducks suffer from fewer diseases than chickens, which should be vaccinated against Marek's and Gumboro, and which also pick up coughs and sneezes from poultry shows. None of these ailments afflicts waterfowl. Newcastle disease, which occasionally visits this country, affects poultry badly but does not kill ducks. Coccidiosis, which is often a problem in rearing poultry, should not occur with ducklings.
- External parasites are also less of a problem in ducks than hens. Given good conditions, ducks should remain relatively free of northern mite and lice, and do not get scaly leg mite.
- Housing is cheaper for ducks than poultry. Duck houses are simple affairs – basically a secure, well-ventilated box, without any complicated nest boxes and perches. Ducks are happy to lay their eggs on the floor, preferably in clean litter, and most breeds will oblige by laying before 9.00am.
- Unlike chickens, ducks can be herded and driven when you want them to go in.
- Ducks do not mind wet, cold weather if they are fit, and require less shelter.
- Several drakes may be kept in a flock together, as long as there are plenty of ducks. The drakes do not usually fight (unlike cocks, which may need to be kept separate from each other).

Ducks are Economical

- The small-scale production of duck eggs does not demand that the birds are tested for salmonella before the eggs are sold, and duck eggs generally command a higher price than hens'.

- A free-range Khaki Campbell will beat a free-range chicken for egg-laying capacity, though probably not an intensively housed battery hen. The ducks themselves stay productive up to four years old, when their laying capacity does decline.
- Ducks do need more space than poultry, and will also consume more food. If they have free range, however, they will find some of their own rations from eating grass and weeds, as well as insect and animal protein.
- Ducks will eat left-overs from the vegetable garden and the kitchen. They particularly like chopped greens and can be trained to eat other vegetables if these are chopped, or softened by boiling.

Ducks Make Good Pets

- Ducks are never aggressive with people. A cockerel can be a thorough nuisance, flying up and displaying his spurs. Drakes do not behave like this.
- As with all hand-reared poultry, ducks can be very tame. Call Ducks in particular make excellent pets.
- Ducks are fun. As well as the interest of hatching, rearing and sometimes showing pure breeds, the birds can be a source of great pleasure as tame garden pets. They are very relaxing to be with after a stressful job.

Ducks are Useful on the Farm and in the Garden

- Ducks help in the garden by consuming pests, especially slugs and insects. William Cook (of Orpington) recommended feeding the ducks green vegetables in their runs and then allowing them out to forage in the kitchen garden for half an hour at dawn. Their sensitive beaks find slugs, which are difficult to see. However, if they have continuous access to the garden they also like to eat many of the crops, such as blackcurrants, lettuces, young cabbages, and so on. Be prepared to fence them off from certain areas when the crops are vulnerable.
- Ducks have long been used on farms for controlling liver fluke. By eating the snails, which live in marshy places, the ducks break the cycle because the snail harbours a life stage of the parasite.
- Ducks can also be used for sheep-dog training. They react very quickly to being rounded-up, and exhibitions of ducks and dogs are sometimes put on at agricultural shows. Ducks in lay are not suitable for this activity – only the drakes should be used. Campbells and Indian Runners (not the tall exhibition varieties which might go 'off their legs') are the favourites.

Ducks for Exhibition Make a Rewarding Hobby

- Ducks are easy to show compared with some poultry. As long as they are kept in good conditions with plenty of clean water, they wash themselves.
- If you are interested in ducks for exhibition, you will find out about the breed characteristics much more quickly through comparing stock at shows and studying breed standards.
- Exhibiting waterfowl can be done at small shows at the local level, mainly at agricultural shows. This need not be expensive.
- Specialist shows with 300–900 waterfowl are

Fig 121 Indian Runners at the Devon and Cornwall Waterfowl Show, Wadebridge, 1999.

spread at several locations around the UK. Ask the British Waterfowl Association for details. These shows are a good opportunity to meet other enthusiasts.
- All the family can join in; there is a junior section too at many of the larger shows.

Considerations Before Buying Stock

Find out about duck behaviour first. Domestic birds are not far removed from the Wild Mallard so, in caring for domestic ducks, the habits of the wild relative should be observed and similar conditions given to the domestics wherever possible. Also, consider if you are at home at the right time to look after the ducks. What happens when you are on holiday? And what arrangements can you make when you are back late and cannot shut them up?

- Ducks are not time-consuming to look after, but the basic requirement is that they are shut up before dusk each night. They will be lost to foxes or other predators, unless they are in a fox-proof pen.
- Ducks need space. Do not over-stock. Start off in a small way to find out how many ducks the land can cope with. Ducks can quickly make a mess on heavy clay soils by dibbling for worms, but free-draining sandy soils can happily accommodate larger numbers. The ideal number of birds kept will depend mainly upon wet-weather facilities, particularly if there is only a small area. Gardens and pens designed with some gravel and concrete areas for winter will keep the place tidy, but this may affect the quality of the birds' feet. Webs have been designed for walking on soft surfaces and can become callused and sore on hard ground.
- Ducks do need water for washing and preening but ponds and streams are not essential if there are suitable-sized water containers. If there are ponds and streams, do not over-stock. Any livestock can cause pollution problems if the numbers are too high.
- Find out about the different breeds. They have been designed for different purposes and do vary in their habits and how much food they consume. Many people want pure breeds for their looks, even if they do not want to show the ducks. There is added interest in keeping and perhaps breeding them. Also, they do have their distinctive temperaments; an Indian Runner is quite different in behaviour from a Call.
- Birds need to kept in the right combinations of size and sex to keep them happy. Find out about breed behaviour and proportions of ducks and drakes.
- Always start with just a few birds to see how any neighbours get on with them first. Call Ducks can be noisy and a source of annoyance. Drakes are much quieter than ducks and can be kept as unobtrusive pets.
- The ducks may be blamed as the reason for rat infestation. Ducks are untidy eaters and it is best to develop methods of feeding which minimize waste.

Buying Stock

Getting to Know the Breeds

The majority of people are not aware of the great variety of ducks breeds. Many people have heard of the Wild Mallard and perhaps the commercial Aylesbury and Khaki Campbell, but that is usually as far as it goes. More information about the breeds can be found in the Waterfowl Standards, where the size, shape and colour of the birds are described in detail, but the best experience is first hand, at a duck breeder's or at a show.

Whilst getting to know the breeds, decide which is most suited to your situation on both practical and aesthetic grounds. Small ducks are best for small gardens, but with good surface materials and a concrete-lined pond, people do successfully manage heavy ducks and Runners in small areas.

When and Where to Buy

Hatching Eggs

You may wish to start off by hatching some eggs yourself in an incubator or under a broody hen. Such eggs are generally available in spring. This may seem a cheap way of starting off but, unless you have done this before, it may not turn out to be so; a certain amount of luck or expertise is needed for successful hatching.

Hatching eggs command a higher price than eating eggs but there is no guarantee on how they have been stored, how old they are or if they are fertile. If they do not hatch, the supplier cannot be held to be any under obligation because it may be due to management technique that the eggs have failed.

Although the eggs have been sold perhaps as pure-breed eggs, it is a long wait to see if the ducklings are good examples of the breed when they have grown up. It is often better to start off with a good pair of pure-bred adults.

Breed summary

Breed	First Year Poultry Club Standard	Purpose of Duck	Weight (lb): Male/Female	Egg-laying Capability*
Heavy Duck				
Aylesbury	1865	originally table, now exhibition	10–12/9–11	35–100
Blue Swedish	1982	ditto	8/7	120–150
Cayuga	1901	ditto	8/7	100–175
Pekin	1901	ditto	9/8	100–160
Rouen	1865	ditto	10–12/9–11	35–100
Rouen Clair	1982	table and exhibition	7½–9/6½–7½	170–200
Saxony	1982	table and exhibition	8/7	150
Silver Appleyard	1982	originally eggs and table	8–9/7–8	150
Light Duck				
Abacot Ranger	1997	egg-layer	5½–6/5–5½	up to 300
Bali	1930	egg-layer, ornamental and exhibition	5/4	150–250
Campbell	1926	egg-layer	5–5½/4½–5	can lay over 300
Crested	1910	ornamental and dual purpose	7/6	100–175
Hook Bill	1997	egg-layer	5–5½/3¾–5	not known; 'everyday layer'
Magpie	1926	general purpose egg-layer	5½–7/4½–6	125–225
Orpington	1910	general purpose	5–7½/5–7	150–200
Welsh Harlequin	1997	egg-layer	5–5½/4½–5	up to 300
Indian Runner (nine standard colours)				
Indian Runner	1901	egg-layer	3½–5/3–4½	140–180
Bantam Ducks				
Black East Indian	1865	ornamental	2/1½–1¾	25–125
Crested Miniature	1997	ditto	2½/2	not measured
Silver Appleyard Miniature	1997	ditto	3/2½	80
Silver Bantam	1982	ditto	2/1¾	60
Call Ducks (nine standard colours)				
Call	1865	ditto	1¼–1½/1–1¼	0–80

* Egg-laying capability is an estimate only. The number of eggs is as dependent on the strain as the breed. It also depends on the age of the duck. Estimates are for yearling ducks.

Commercial Stock
Commercial stock can be bought as ducklings, or at different growth stages up to point of lay ducks (twenty-four weeks old). Khaki Campbells can be bought as sexed day-olds because this is a laying breed and the hatcheries therefore dispose of the males. White table ducks can also be purchased, sometimes in the supplier's own strain. Most ducklings are available in spring between April and June but large suppliers have a much longer production period. Before buying 'out of season', consider that ducklings will need more care and housing if reared over winter months.

Some markets still accept ducklings for sale, but this is not good practice as the birds are often there when the weather is cold in spring. The ducklings become chilled very quickly unless they are sold with a mother duck or foster hen. If a number of commercial ducklings are needed, they are probably better bought from a reliable supplier who advertises regularly in a journal such as *Poultry World*. When buying ducklings, always make sure that all the facilities are ready for rearing them.

Pure Breeds
A visit to a show often sparks off the desire to keep pure breeds. Shows generally provide a catalogue that lists the breeders, so this is one way of finding out who has the type of duck required. These breeders, and other suppliers who may not show, often advertise in specialist journals such as *Country Smallholding* and *Smallholder*.

As a general rule, it is the breeders who show who will have the better quality ducks for sale, especially in the pure breeds. The larger enterprises tend to supply utility strains in larger numbers, and more cheaply, because they are generally concerned with mass production and strains that are easy to hatch.

Sales and auctions, which include pure breeds of waterfowl, have become more common in the 1990s but vary a great deal in the quality of stock available. The annual Rare Breeds Sale, when it ran at Stoneleigh, offered only graded pure breeds for sale. At other auctions, buyers have to take the vendor's word for what is on sale. That is why it is advisable to have some knowledge of the breeds before buying.

The better sales and auctions generally offer only adult birds for sale. This is also the case with breeders who will generally not sell birds until at least in their first feathers at eight to ten weeks old. Birds like this are generally not available until July, and most adult pure breeds are sold between August and November. This is when there is the best choice in quality stock. As the winter progresses, breeders cannot hold large numbers of stock due to cost of feeding and poor weather, and only the best birds will be retained. Consequently these will be more expensive and in shorter supply.

Sex

When buying any livestock, examine it carefully. First of all check that the birds are the sex required. Ducks quack; they have a loud voice. Drakes are often quieter when handled, but give a lower, husky noise when they do speak. There should be a definite difference between the two sexes by five weeks old.

In coloured breeds, there are often pronounced differences between the sexes, even at six weeks old, and breeders should be able to sex coloured breeds by both bill and feather colour at this point. By sixteen weeks, drakes are getting the characteristic Mallard curled sex-feathers on the rump. These can, however, be removed by wear and tear. Always confirm sex by voice or colour of plumage and bill. Ducklings can be vent-sexed; this is how commercial layers are sorted out as day-olds.

Do not buy more drakes than ducks, unless only drakes are wanted. Drakes that have been brought up together can live amicably with no problems but a surplus of drakes will cause havoc with laying ducks because of repeated mating. Drakes that are a particular nuisance in the breeding season must be removed from the duck pens.

There are usually surplus drakes that can be sold as pets. However, pure breeds are usually sold in pairs (a duck and a drake). It is unreasonable to expect to buy large quantities of females from a breeder because pure breeds are not reared in large numbers, and drakes are reared for breeding and exhibition purposes. If only females are wanted, in large numbers, find a commercial supplier where the drakes are culled at day-old.

Health

Check the health of the birds by handling. They should be caught quietly by first driving them into their usual shed in a small group; this avoids a panic because large groups of frightened birds will pile up in a corner and injure themselves. Birds should be caught using both hands; they may be restrained by the wing or neck but the body weight must be supported when lifted. Make sure they are heavy enough for the breed. Birds can be light because of under-feeding but can also be ill. Adult birds should have a plump breast that almost conceals the breast bone. Check that the bird's eyes are clear and do not have an opaque glaze. Also check the under-sides of the bird's feet for calluses. These can come with age or poor conditions.

Watch the birds moving. They should be busy and alert. Birds that are too easy to catch are probably ill. The ducks should move comfortably; watch out for deformities such as crooked neck, roach back and wry tail. The wings should also be set snugly to the

Fig 122 Never catch a duck by the legs. Restrain the bird by the neck and slide the right hand under the body to hold one thigh and catch the other leg in the little finger. Keeping two or three fingers between the bird's legs prevents squeezing them together too tightly. A duck's legs and hip joints are off-set at the sides of the body for paddling; pulling the legs together puts too much pressure on these joints. Lift the bird up using your other hand against the bird's side, while also catching hold of the wing. Tuck the bird against you body, its head and neck under your arm.

Fig 123 The underside of the webs and toes should be soft and free from blemishes.

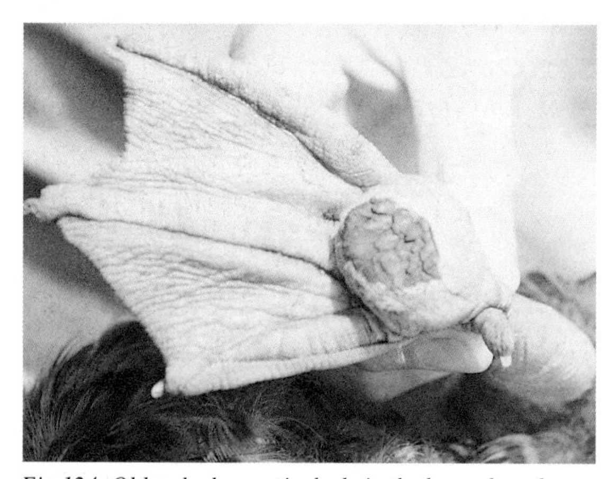

Fig 124 Older ducks, particularly in the heavy breeds, may develop a hard lump under the heel of the foot. This is known as bumblefoot and is difficult to clear up. It may not bother the bird if it is kept on grass and has access to swimming water.

Fig 125 Indian Runners tend to develop thickened skin at the joints on the underside of the feet. This may be to do with their upright stance, and the fact that they seem to spend less time on water than other ducks.

body, not dropping down or twisting out. Finally, check the plumage. The vent should be clean and free from droppings; a soiled vent indicates disease. A healthy duck is clean to pick up, with tight, glossy feathers.

Age

Ducks do live up to ten years old, and even longer in good conditions. Healthy females are good layers for up to four years but after that the number of eggs will decline significantly. Also the eggs can become misshapen and thinner shelled. Young ducks are the most productive but good specimens for breeding can be up to four years old and the best ducks can be kept for several years; the quantity they provide does not matter if the quality is good. Two-year-old ducks are preferable to yearlings for breeding quality birds in Runners and the heavy breeds. The ducklings grow better from the slightly larger eggs.

Occasionally, ducks change sex as they get older, i.e. females acquire male characteristics. The deep quack becomes more of a hoarse whisper, the head and rump plumage becomes more masculine (in coloured birds) and sex curls can develop on the rump. No further eggs are laid. This can happen as early as two or three years old, probably in several breeds but certainly in the Calls, Saxony and Orpington. It may be ovary damage or a sign of too much in-breeding – and definitely a dead end.

Fertility in drakes depends upon the breed. In the heavier breeds, young, fit drakes are preferable to heavy old birds. In the lightweight Runners, Magpies and Campbells, drakes can still be fertile at seven years old. It is therefore not necessary to buy a

Table from Powell-Owen (1918) to show how egg production declined over four years. In 1918 there were no official laying test records in Britain, so Powell-Owen used results from Hawkesbury Agricultural College, New South Wales, where six ducks comprised each pen. The average (in brackets) is per bird.

Owner and Breed	First Year	Second Year	Third Year	Fourth Year
1. G. E. O. Craft: Indian Runners	1,034 (172)	1,019 (170)	826 (138)	762 (127)
2. D. Salter: Indian Runners	1,092 (182)	1,075 (179)	644 (107)	626 (104)
3. Hughes Bros.: Buff Orpingtons	1,105 (184)	882 (147)	703 (117)	625 (104)
4. G. Rogers: Indian Runners	1,220 (203)	1,244 (207)	719 (120)	588 (98)

NB The Buff Orpington was a relatively new breed at that time. Orpingtons today, and exhibition Runners, do not necessarily produce these numbers of eggs. The strain is as important as the breed for production.

yearling pair of birds for breeding quality, but aged birds should be avoided.

Transport

Ducks need cool, well-ventilated containers for travelling. The size of the container should match the ducks. Indian Runners ideally need a container that enables them to stand up and stretch their necks; they occasionally develop a permanent kink in the neck if they have been constrained by a low box.

Pet carriers made of plastic, with ventilated sides and a wire door are ideal but, if they are not available, cardboard boxes are adequate. These must be prepared by cutting several 6×4cm holes to allow free-flow of air to keep the ducks cool. If several boxes are packed, care should be taken to ensure a flow of air between the boxes; there is no point in cutting air holes if these are then obstructed. On no account should cardboard boxes be stacked, as the cardboard can collapse. Also, the boxes should not be free to move in the vehicle.

The best transporting boxes are rigid plastic poultry crates that are designed for stacking and for free-flow of air. These are ideal for hot weather when particular care is needed in transporting any poultry. All poultry can soon die from over-heating; their feathers make them particularly vulnerable inside an insulated box. Vehicles should always be kept cool, in the shade, before loading. There must also be a flow of air in the vehicle whilst travelling; professional carriers use well-ventilated vans. The boxes must not be confined to a sealed boot of a car. Choose suitable litter to mop up any mess. Straw gives a good foothold on a slippery plastic surface, and wood shavings on top absorb droppings. Food and water are unnecessary for short journeys. Regulations over this apply to commercial transport of livestock for journey times of over eight hours.

Fig 127 Poultry crates and pet carriers, which are well ventilated, are best for carrying ducks, especially in hot weather: Call Ducks and Abacot Ranger drakes in different crates, suited to their size.

5 Managing Adult Stock

Before buying new stock, it is essential to prepare their new premises. It comes as a bit of a culture shock to birds reared with a flock of ducks to arrive at a place where there may be only two. Apart from the breeding season, Mallards are naturally flock birds. The first response of birds moved from a flock is to try to find some more birds, so they need restraining from running or flying off.

Ducks are timid birds and need to be handled quietly and calmly. When they arrive at a new place they should not be confronted with unfamiliar animals and they will need time to get used to their new owners too. Ducks know their regular handlers, so do not expect them to be tame straight away.

Arrival and Unpacking

The birds should be released in a secure area. The best place is their new shed. They can be left for a few hours to settle down, with food and water, before they are let out into a wired area so that they cannot escape. Some breeds such an Indian Runners and Black East Indians can be upset and run off, whereas the heaviest breeds such as the Aylesbury and Rouen are not capable of moving so fast.

It is particularly important to fence in new ducks if there is a stream. A river is a route to better things for ducks, and they need to have identified with the new place before they sample a river. Get the birds used to feeding at a certain time at their shed before giving them their liberty. Remember that rivers in flood are a hazard and will sweep stock away. Keep the duck house away from the river too, above the flat land that will flood, and pen the ducks in a safe place when the water is high.

Clipping

Breeds that are capable of flight should be clipped on arrival, if not at the point of sale. All of the Bantam Ducks and Calls come into this category. Although the light breeds, such as Welsh Harlequin, should not fly, a few determined individuals can and it is better to be safe than sorry. This is especially important if the bird is a bit light. Indian Runners and the heavy breeds cannot fly.

Clipping involves cutting four to six primary feathers on just one wing. This unbalances the bird when it tries to fly, and puts it off. The primaries are cut with a pair of scissors where the coverts just overlie them, through the solid stem (rachis). This does not hurt or harm the bird, and the primary feathers moult and grow again the following summer.

Fig 128 Clip the wing feathers of the smaller ducks that are able to fly. Using scissors, cut five flight feathers where the coverts overlie them. This is only done on one wing.

Housing

Ducks can live outside in most weather as long as a shelter of trees and bushes is provided. However, they must be protected from vermin. Ducks do not roost; they live on the ground or on water for protection in the wild. Most of their predators are nocturnal and so domestic ducks will not last long if they are left out unprotected overnight. It is no use expecting them to take refuge on a pond or island. Domestic ducks do not know about predators and some predators, such as mink, are aquatic too.

Unless there is a fox-proof pen, ducks must be guided into a shed each night with outstretched arms or a long stick. After a couple of evenings, the birds know where to go and it is only a minute's job to direct them in each evening.

Design and Materials

Duck sheds are designed with a wider door than a pop-hole suitable for chickens. Hens are happy to move through confined spaces on their own, without being driven. Ducks like wide doorways to move through as a flock. They dislike going in through narrow openings, and tread all over each other trying to spill out of them in the morning. The opening should be at ground level.

The shed must be dry and well ventilated. A weld-mesh panel situated high up, rather than at ground level, prevents drafts on the floor as well as vermin disturbing the birds.

Commercial roofing material is usually wood covered with roofing felt, but corrugated sheeting can be used. Metal types tend to drip with condensation. If plastic is used, it should be opaque; ducks do not like shadows on moonlit nights moving across their shed. The corrugated material Onduline, available at larger DIY stores, has good insulation properties and has the advantage of being a single layer so that pests such as red mite are not harboured as they can be under roofing felt. However, make sure that the material has sufficient strength to provide protection from determined predators.

The housing materials should take account of animals in your area that might attack ducks. Badgers are very strong, and can pull wooden and wire structures apart. In these circumstances, ducks are better kept in a brick and concrete outhouse or garage. New wood, treated with preservatives for a longer life, is generally effective against foxes, polecat and mink. Old, rotting wood is no use; rats can gnaw their way in very quickly and have been known to kill small Call Ducks confined in sheds. The base of the shed is particularly important – it needs to be solid new wood, to exclude rats. Place the shed on concrete slabs to make the base last longer on a dry site. For large numbers of ducks, concrete floors are best, but these are cold and need to be covered in a deeper layer of litter.

Wooden tongue and groove sheds are usually sold treated with preservative, but this should be renewed annually on the outside only. The ducks must be rehoused when this is done and the fumes allowed to disperse. Treatment with creosote might require a month or more before the accommodation can be used again. Read the instructions on using preservatives, and use your own judgement.

Always go for the best materials that will last, rather than for eye-catching designs. Second-hand sheds are often not much use: the materials may not last much longer and they may harbour red mite from poultry.

Sheds can be obtained from garden centres, listed in the *Yellow Pages*, and smallholder magazines carry advertisements for suppliers of poultry housing.

Site

The siting of the shed is important. It should be next to a fence where the ducks can be directed into the door; if the shed is free-standing, they will run around it. The door and ventilation panels should face away from the prevailing wind. The shed itself must be placed on well-drained land that will not flood. If the shed is in a very sunny place the ducks must be let out early on hot days in summer; they will overheat and die if left in a small wooden shed.

Size

The floor space per bird will depend upon the breed of duck. Heavy ducks such as Aylesbury and Rouen,

Fig 129 A simple well-ventilated shed designed as a dog house is useful for ducks.

and Indian Runners which are long in the body, need 3–4sq ft per bird. So a 4ft square shed would house four large ducks. It is important not to overcrowd in the summer months when temperatures are high. The shed roof should not be too low; a minimum height of 2½ft (76cm) is needed for sufficient ventilation.

An Additional Run

It is not always possible to be at home when it is time for the birds to be shut up. In these circumstances, it is useful to have an additional vermin-proof run to let the birds onto for feeding and bathing. This can be combined with a wet-weather surface for the ducks (*see* page 125).

Bedding and Cleaning

Peat moss and bracken are often recommended as bedding, but the most commonly available materials are straw, wood shavings and sawdust. To be effective, the straw must be chopped. It is more absorbent and easier to handle. Plastic-wrapped bales of dust-extracted straw are convenient but up to four times more expensive than farm-baled material. The cost is similar to wood shavings, and it would be worth experimenting to see which suits your system better. Straw is better for the compost heap than shavings.

Rough sawdust is the best material. It is cheaper, very absorbent and reduces smells. It must be free from pesticides and preservatives, and be coarse, not powdery. It need not be removed from a brick or concrete floor, and can be built up into a deep layer over the winter, providing good insulation for the birds. This can also be done with straw but it tends to hold the moisture more so that ammonia builds up in the shed and moulds develop. Deep straw works better on an earth floor, but weldmesh beneath the straw to exclude rats is essential.

Fig 130 A house made out of tanalized tongue and grooved wood. It has a variable ventilation panel situated high up and a wide door. The pop-hole is all right for use once the ducks have become familiar with it. The nest boxes for chickens (on the left-hand side) are unnecessary for ducks.

In wooden-based sheds the droppings and litter can be removed each week, rather than let the litter build up. In either case, the ducks will have a preferred area for laying and this needs to be topped up each day with clean shavings. Encourage this with an easy step-in box or secluded corner at floor level.

Hemp is available as a bedding material for horses but is unsuitable for adult ducks because, if they eat it, it swells up several times with added moisture. This will kill ducklings.

For top egg production, ducks probably do need overnight access to water. This used to be supplied on slatted verandas to minimize mess in beautifully designed commercial systems in the 1920s and 1930s. This is not necessary for most ducks. If water is allowed in a bucket in the shed, the birds put a mess in the bucket, and make the bedding wet and unpleasant. They are best left without food and water overnight, but must be let out early on summer mornings when the weather is hot.

Water

Ducks are water birds but they can manage without water for swimming. Indian Runners are least bothered and Call Ducks probably like it the most. However, all ducks must have clean water for bathing their heads frequently. This keeps their eyes clean; they get sticky eyes and eye infections if water is insufficient. Water is also needed to keep feathers in good condition and the birds must be able to use their head and neck to throw water over all their plumage and use their preen gland. This is why Call Ducks need to swim because it is hard for them to duck their short necks into a bowl for washing; it is better for them to get in it.

Natural Water

Running water is best because it does not freeze in winter and it stays clean. Free access to this is not advised until new ducks have become familiar with the place. A small stream is better than a river; rivers in flood can sweep birds away and they can also harbour mink.

Ducks keep natural ponds clear of weed and the birds will find a lot of their own high-protein food. Small ponds with decaying vegetation can become a hazard in a hot summer where they may become a source of botulism.

Fig 131 If the stream is small, ponds can be constructed by making small concrete dams. The ducks keep the ponds clear of sediment by their activity. The water does not freeze in winter. In a fox-proof pen, this is the favourite place to spend the night in freezing weather.

Fig 132 A series of small concrete pools can be constructed on a sloping site. These are connected by a concrete channel so that dirty water drains naturally from the area. They can be filled on a continuous basis, or brushed out and refilled as often as necessary. (Call Ducks at Steph Mansell's)

Artificial Pools

If a concrete pond is made, make sure it can be emptied and refilled easily, preferably with a drain (sump) emptying into a ditch or sewer. On sloping ground, a series of ponds connected by a chute can be filled by a hose pipe at the uphill end and then drain naturally into a ditch or soakaway. This system requires least maintenance and can be kept ice-free by running water through it continuously.

Ponds can be made by sinking a polythene liner into a hole in the ground, but these are eventually ripped by the ducks' claws, and soon become fouled by droppings. It is better to buy a fibreglass paddling pool to empty and refill on a regular basis. In freezing weather, the pool should be emptied and turned over so that ice is not a hazard. The pool is best placed by a drain, or free-draining area, so there is no problem in water disposal.

Smooth-sided ponds can be hazardous for ducks. If the pond is full, the birds just walk or flap out. It the water level is low, young birds cannot get out. They try many times before they get water-logged and exhausted and then drown. With ducklings, always make sure that they can get out of water containers, e.g. by placing a rough brick as a stepping stone inside a bowl. One place at a pool's edge is not enough; they may not find it.

Buckets and Bowls

Ducks can manage with just bowls of water and some breeds, like Indian Runners, prefer to spend more time foraging in the grass than on water. If their water is from bowls, all ducks must have clean water at least once a day. Demand is much higher in the summer, and the bucket should be placed in the shade to keep the water cooler and bacterial and

Fig 133 Placing the water container on a weldmesh surface (which prevents the ducks from dibbling) keeps the water cleaner and the birds in healthy conditions. The weldmesh can be raised on bricks if the ground is very soft, or placed over a container to catch spillage.

algal growth down. The bucket should be scrubbed once every few days as the inside becomes slimy.

Although a bucket will do for drinking water, it will not keep ducks in the peak condition needed for shows. There is no substitute for lashings of clean water to give black ducks a terrific sheen, and to keep Pekins white and fluffy.

If swimming water is not available, care must be also taken over the ducks' feet. The heavy breeds need soft or grassy land because they can develop callused feet. Runners are also prone to calluses developing at the joints under the toes. This is probably as much to do with stress on their feet from their upright posture and active character. Other ducks kept in the same conditions often have perfectly smooth-skinned toes.

- Ponds and buckets are good for ducks but hazards for small children. Fence young children out of the duck area.
- Never leave ducks without water, especially if there are food pellets about. The ducks eat these, then they swell up inside the bird when it drinks some time later. This will kill ducklings, which prefer to eat and drink simultaneously.

Adult Diet and Feeding

This section is only about feeding adult ducks outside the main breeding season. This means from July or August until January. During this period birds should be either on a maintenance or a layers' diet, depending on the purpose of the duck.

Ducks are omnivores, and if adults free-range in small numbers they will find most of their food. Supplementary feeding is still desirable, however, to check that they are getting enough food, to increase egg production if the ducks are laying varieties and also to bring the ducks home in the evening. If the birds are on free-range, always feed them in the same place and time (an hour before dusk) to establish regular habits.

It is often more convenient to restrict the ducks' range and feed them largely on commercially produced pellets and whole grains. Pelleted food in Britain is generally manufactured for laying hens but duck food is available; telephone the major feed manufacturers to find your nearest depot.

Pelleted Food: Ingredients

Pelleted food is made from milled ingredients, which are mixed with an agent to make them stick together before they are extruded under pressure. Meal or mash is made from a similar material to the pellets but it is dry and powdery and not in a suitable form for ducks – it sticks in their mouths. The meal can be mixed into a soft, crumbly mixture with water but is more time-consuming to feed.

The bulk of pelleted food is provided by milled wheat. This is either whole grain or middlings, which are a by-product of the milling industry (producing white flour) and quite high in protein. Whole wheat itself varies in protein from 8–12 per cent depending on the variety and climate where it is grown. In some pelleted foods, by-products of pasta are used as a way of using up ingredients that would otherwise be waste but are basically wheat.

Fish meal was once used to enhance the protein and mineral content of poultry foods. However, it is no longer so easily available and has probably become more costly to obtain. This is unfortunate as fishmeal contains many of the essential vitamins and minerals for a healthy duck's diet.

Meat scraps were always recommended in old poultry books to improve laying, as well as growth of ducklings. Meat meal was added to many animal feeds up to the 1980s. It was a product of the meat-rendering industry, where animal products unfit for direct human consumption were processed for use in animal feeds. However, pellets no longer contain meat residues, which were banned from British manufactured animal feeds after the BSE crisis of the 1980s and early 1990s. The 16 per cent useable protein content of layers' pellets is achieved by using soya bean as well as peas, beans and oil seeds, such as sunflower, linseed and rape. At present, genetically modified soya is almost certainly in manufactured animal feed unless the company supplying the feed declares the feed to be GM-free. Most animal-feed soya is imported from the USA, where there has been less public resistance to GM foods.

Hen layers' pellets also often contain an ingredient such as canthaxanthin to enhance the colour of the egg yolk, which is not necessary in free-range birds. The ingredient can be marigold extract. Read labels carefully if unnecessary ingredients are to be avoided.

Vitamins and minerals are added to all pellet foods in the manufacturing process. Genuinely free-range birds are able to pick up trace elements from wild protein and greens but added iron, zinc, molybdenum, copper, selenium, iodine and sodium are all beneficial. Selenium is deficient in British wheat compared with North American and so is usually added.

Limestone flour (ground limestone) is included to increase the calcium content. In addition di-calcium phosphate (phosphate of lime) is provided as a supplement because the correct ratio of phosphorus to calcium is needed for the ducks' metabolism. Di-calcium phosphate was manufactured from animal bones, so all feed companies should now be using added calcium and phosphate derived from rocks. A local feed company, for example, adds ground rock imported from Tunisia.

Calcium and phosphorus are both essential for bone structure and egg quality, but they must be in the correct proportion. A higher proportion of calcium is needed in layers' feed than growers'. Hen layers' rations typically contain 4 per cent calcium needed for eggshell production but Laing (1999) notes that in poultry, this will cause kidney failure if fed to non-laying birds, which need only 1 per cent.

Pelleted Food: Suppliers

Local food mills usually give the best deal on prices, but will probably offer a limited range of products compared with the larger national companies, which have become much more duck-food conscious. Check the *Yellow Pages* for the area under Animal Feedstuffs to find out what is available and compare the list of ingredients on the labels on the bags with proper duck food before making a choice.

Companies such as Bibby's (Slimbridge), BOCM-Pauls Ltd (Marsdens Game Feeds), Marriages, and Allen & Page now produce a variety of rations suitable for waterfowl at different stages of growth and times of the year. It is possible to buy pellets that are free from yolk pigment and which contain the correct amount of vitamins and minerals for adult birds. These can be purchased as layers' pellets (16–17 per cent protein) or as a maintenance diet (14 per cent protein) for birds that are not in lay and that also need less calcium.

Feeding and Storing Pelleted Food

Adult ducks should not be given a diet of just layers' pellets, especially if they are kept as a mixed flock of ducks and drakes, and some birds are in lay and others are not. Egg-producers want to eat the high protein/high calcium pellets. Drakes and non-layers, given the choice, will balance their own diet to a lower protein and calcium level by consuming more wheat. In practice, many people feed a half-wheat/half-layers' diet because this is simplest.

Pelleted foods have a limited shelf-life during which vitamins deteriorate. Always check the use-by date on the label and examine the food before using it. Badly stored food may develop blue-green mould; this can happen before the food is packaged at the mill, usually if pellets have lodged in bins or hoppers supplying the bags. Once you have bought a sack of pellets, store it in a cool, dry place and use it by the recommended date. Make sure that vermin and wild birds cannot get access to the food. Apart from wastage, rats and mice can carry salmonella, and wild birds can transmit avian tuberculosis and bacterial disease.

Whole Grain

Grain is not a complete food. It needs to be supplemented either by suitable pellets or by food found on free-range. Laing (1999) notes that some breeds of hens, particularly if they lack exercise, suffer from fatty degeneration and haemorrhage of the liver if fed on high-energy, low-protein rations. However, it is beneficial to feed ducks some whole wheat, the quantity varying according to the time of year, as long as they do not become too fat.

Wheat can be fed dry, mixed with the duck pellets. Alternatively the grain alone can be left under water in a bucket. This has the advantage of keeping the food clean of wild bird droppings and also reducing waste from wild birds eating the food. If starlings and sparrows are a nuisance, there is little wastage to continue to attract them. Ducks are messy eaters but comparatively little wheat is spilled in this way. The ducks also like eating the wheat under water; this is a natural way of feeding for them. There is the additional advantage of making the birds wash their eyes if there is little alternative water available. Small ducks need their wheat fed under water in a bowl instead.

Although mixed corn (wheat and cut maize) is often fed to poultry, this is much more expensive than straight wheat and the maize seems of little benefit to the ducks except during cold, winter weather. Maize contains more carbohydrate and fat but is lower in protein and vitamins than wheat which contains up to 12 per cent protein.

Barley is often cheaper than wheat but the whole grains are spiky and not liked by ducks. Like maize, barley is better fed as a fattening diet for market ducks rather than a maintenance grain.

Food Containers

Food should not be thrown on the ground because much of it will be wasted and attract rats. Apart from a plastic bucket for water and wheat, the ducks will also need a trough or plastic container for dry pellets. Make sure that there is enough space for all the ducks to feed, or enough time for them to have their fill. Domestic ducks are greedy eaters and if they are kept with wildfowl these slower eaters can be starved.

Ducks are enthusiastic eaters and grab a beak-full of food to jerk down their throat in a series of lunging movements instead of swallowing in a more controlled way, like a mammal. The hungrier they are, the more food they will spray around them. Although Eltex and pig troughs are useful for large numbers of birds, because they give more feeding space per bird along the line, plastic washing-up bowls are easier to clean and better at catching all the bits of food. If you buy a well-known high-quality plastic, it will last much longer than cheap types, which quickly turn brittle and break in sunlight. Some colours also last longer; green and brown seem better than blue and white.

Never feed new, dry food on top of old, damp food. The damp food will stick to the bottom of the container and start to go mouldy. Scrub the inside of the container if wet food has adhered to it. Move the container to a new patch periodically; pellets spilled on the ground also become smelly, develop moulds and are a health hazard.

Amount and Type of Food

The daily amount varies according to the breed and its purpose, the weather, how the ducks are housed and if the ducks are laying. Birds living outdoors in a fox-proof pen eat far more than birds that are housed. Thear (1998) (for a laying hen) quotes an extra 4.2 calories to keep warm for a fall of 1° in temperature. A laying hen in winter, in addition to a daily ration of 130g of pellets and wheat would need a further 20g of grain. Wheat, in these circumstances, is beneficial in the diet to keep weight on, and give the ducks more calories to get them through cold winter nights.

Fig 134 For large quantities of ducks, a self-feed hopper that can take a sack of pellets is useful, particularly if the ducks are in a fox-proof pen and are to be left for over twenty-four hours. The hood over the circular trough protects the pellets from spoilage in rain. The hopper can also be closed down overnight (galvanized equipment from Bird Stevens and Co Ltd).

Fig 135 Ducklings feeding at a galvanized trough. These give a large amount of feeding space per bird. They can be used for dry food or for wheat under water. They are more difficult to clean than a plastic bowl but last longer.

Heavy ducks eat as much as geese; they can shovel up between 6 and 8oz (170–226g) a day when they are growing. Call and Bantam Ducks eat only a couple of ounces each. Observation of the ducks' habits is therefore necessary to gauge how much they need.

Ducks intended as commercial layers need to be treated differently from pure breed stock ducks, which do not lay all year. Laying ducks are hungry and love layers' pellets, which increase their egg production. They should not be allowed to get too fat, so less wheat should be fed and a higher proportion of pellets given to provide the protein necessary for egg production, but fewer calories so weight is not put on. In laying ducks, internal layers of fat are thought to impede the passage of eggs. Pure breed ducks should be slimmed down for the egg-laying season which, for many, starts about February. Obesity may also result in oviduct, as well as heart and liver, problems.

Many pure breeds, such as exhibition Indian Runners and Rouen, do not usually lay in the autumn. These birds need to put their energy into growing into fit, strong ducks and should be fed maintenance pellets and wheat, and not forced into laying. If the birds are foraging on farm land, they will be finding a lot of animal protein and egg production may start anyway.

Check that you are getting the feeding right by picking up the birds every few weeks. They should be fit but not fat. This means they should have a good covering of muscle on the breast, so that the sternum does not stick out. Indian Runners rarely get too fat because they are very active but the heavier breeds, such as Rouen and Appleyard, do tend to put on more weight.

The Importance of Grit

Birds do not have teeth for grinding up food. Grain and grass must be broken down in the gizzard, so grinding material must be supplied. Provide mixed poultry grit, containing flint (insoluble grit for grinding) plus oyster shell and crushed limestone chips (soluble grit) for calcium for ducks in lay.

The flint does not contain calcium; it is a harder material than limestone or oyster shell, consisting of silicon dioxide (quartz in a microscopic form). The natural material, derived from flint nodules in the chalk, is rather glassy and breaks into sharp-edged pieces. These look as if they might injure the gizzard but Cook (1899) observed that broken flint and even glass (not recommended here) picked up by the ducks eventually passed out as smooth as polished

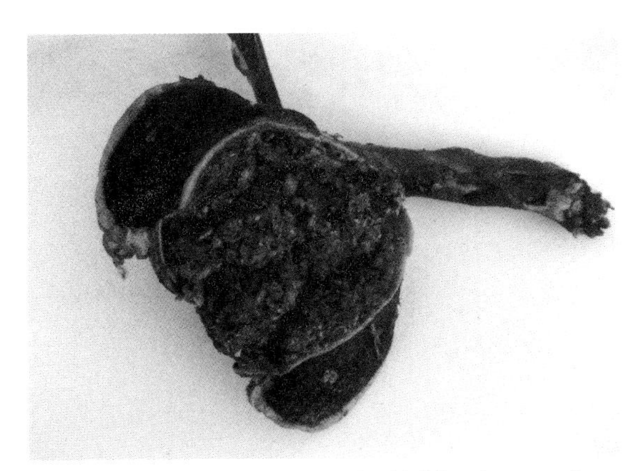

Fig 136 Close-up of the gizzard, which has been cut in half to show the muscular nature of the organ. This duck died accidentally and the gizzard was full of food and grit.

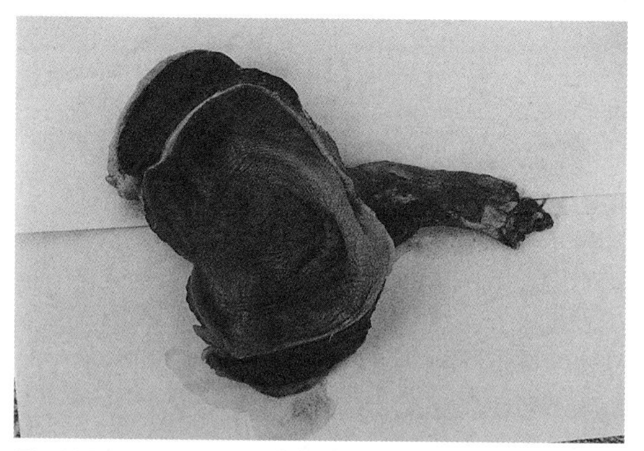

Fig 137 The lining of the gizzard is roughened and becomes horny with usage. Note that the entrance to the gizzard through the proventriculus is larger than the exit.

marble. He thought it was both the grinding mechanical action in the gizzard and the gastric juices dissolving the silica that caused the rounding of the particles.

Grit can simply be left in a depression in the ground but there is less wastage if it is left in a well, formed by a surround of bricks, or in a bowl with drainage holes, e.g. a heavy, broad-based plant pot. For large numbers of ducks, a trough with water running through the grit is ideal. Grit is most important if whole grains are fed; this is particularly so with oats and barley, which have more husk. If the ducks are short of grit, they will pick up anything hard, including nails and staples, which will cause peritonitis.

Quite large pieces of rock may be ingested if grit is in short supply, but these are not as useful for the gizzard as many small bits. Cook observed that a large pebble might even contribute towards blockage of the gizzard because of the lack of grinding action, and even blockage of the smaller outlet from the gizzard by the large lump of stone.

If the ducks are kept in an area where the soil is calcium-rich, on chalk or carboniferous limestone, there may be no need to have mixed poultry grit for extra calcium, but the ducks will still want grinding material. Duck-keepers find that the egg-shells are thick because of the limestone chips, which the ducks find themselves in the soil, and the high calcium content of the grass. The thick egg-shells developed in these conditions may need watching for hatching. In contrast, in lime-poor areas, laying ducks may develop osteoporosis, as well as soft-shelled eggs, if lime is not made available in the diet.

Charcoal

Powell-Owen also recommended charcoal as an essential part of the duck diet. He commented that it 'sweetens the internal organs and is a good purifier'. I have not seen this recommended for ducks elsewhere, but charcoal adsorbs undesirable substances in the gut, and the behaviour of the ducks after a bonfire confirms his recommendation. After the ashes are thoroughly cooled, preferably after rain, ducks and geese relish the pieces of charcoal. If the ash, and particularly the charcoal, is collected it can be rationed out. The bonfire material must be free from nails, staples and toxic materials such as paints and plastics; only garden brushwood and branches should be used.

Greens and Protein

Ducks are definitely healthier if they have free range on pasture. They consume quite a lot of grass and find their own animal protein, which is deficient in manufactured food today. Meat meal and fishmeal were once added to poultry feed. Meat was withdrawn in the 1990s and fishmeal is now rarely added. This may not matter for laying ducks but should be borne in mind when approaching the breeding season, when animal protein found on free range may improve the quality of the hatching egg.

Land: Space Requirements

One cannot be dogmatic about the space required for keeping ducks because this depends on so many variables. It depends upon the season, the weather, soil texture, the breed of ducks, the water supply and one's tolerance about the looks of a place.

The ideal ground for keeping ducks is free-draining, sandy soil, where water rarely lodges on the surface. This allows the droppings to be broken down rapidly by aerobic bacteria and the ducks are discouraged from dibbling for worms because the ground remains firm. It is easier to maintain a turf in these conditions.

The worst conditions are a heavy clay where water will not infiltrate but lies on the surface allowing the ducks to puddle the top layer and restrict percolation of water even further. Stocking must be much lighter, and strategies adopted to prevent the birds making the whole place a mud-bath.

Winter is the worst season because rainfall is generally heavier, the grass is not growing so quickly to utilize the moisture, and evapotranspiration rates are low. Although ducks can puddle ground in a wet spell in summer, it usually dries rapidly and the grass repairs itself quickly. Indian Runners and light ducks in lay are probably the worst birds for puddling the ground because they are enthusiastic foragers for worms, and laying ducks have a large appetite. In contrast, Call Ducks are not nearly so bad.

If there is a large number of ducks it is best to stable them indoors in a particularly wet spell. Another strategy is to confine them to a small area, which they can make as muddy as they like – as long as they have clean bathing water to keep them in good condition.

Another option is to provide an outdoor free-draining surface, especially for wet spells in winter. This could be a sloping concrete yard on a farm. This solid surface is not suitable for long periods because of the effect on the birds' feet. Calluses will develop if they are kept on hard surfaces too long. Yards for prolonged use should be covered in bark peelings, sand or straw, and should also be sloping so that dirty water drains off. These are really only suitable for producing ducks with a limited life, and need farm machinery to handle the larger quantities of materials.

On a smaller scale, gravel surfaces in gardens work well. If the gravel is deep enough the ducks cannot puddle the mud and the droppings are washed down the pores. The best kind of gravel is well rounded, large pieces, too large for the ducks to swallow and also smooth so their feet are not cut. This kind of area need not be large if space is at a premium, and it can also be made into a fox-proof run for days when the ducks cannot be shut up before dusk.

Depending on the preparation of the ground, ducks can be kept at a high density with gravel surfaces and running water to wash droppings away. If only pasture and buckets of water are available then try stocking two or three ducks in an area of 10×10m. Start off in small way to see how the land behaves first under a variety of weather conditions. Sandy areas will have no problems; clays may fare badly.

Fencing: Keeping the Ducks In

Fencing to keep ducks in needs only be made of light materials. Partitions between breeding sets can be made from rabbit wire, preferably 18-gauge, which will last longer. Pig-netting is heavier than needed and the ducks can generally get through the holes. If rabbit wire is folded out by 6–8in (15–20cm) at the base, grass will grow through it and prevent the smaller breeds getting under the wire. Three feet of wire is usually sufficient to stop all but the most determined or frightened ducks getting out. Clip good fliers in the breeding season to keep them in. Miniature Appleyards are the most difficult to keep in place as they will run up and over wire if they want to get back to their original quarters.

Fencing: Keeping Predators Out

Fencing intended to keep predators out has to be more specialized. Where foxes are cheeky enough to hunt in the daytime, appropriate fencing can give peace of mind for the owner and the ducks. Six-foot fencing with a floppy-top or electric wire is needed to stop determined foxes from running up and over. A big fox can get over a 5½ft (1.6m) gate with a rigid top. An electric wire near the base of the fence may also be needed to stop them digging on sandy soils and will certainly be needed if there are badgers in the area. In addition, electric protection close to the wire may be needed to stop climbers. This taller fencing is difficult to put up and is probably best done by a contractor. See examples of different

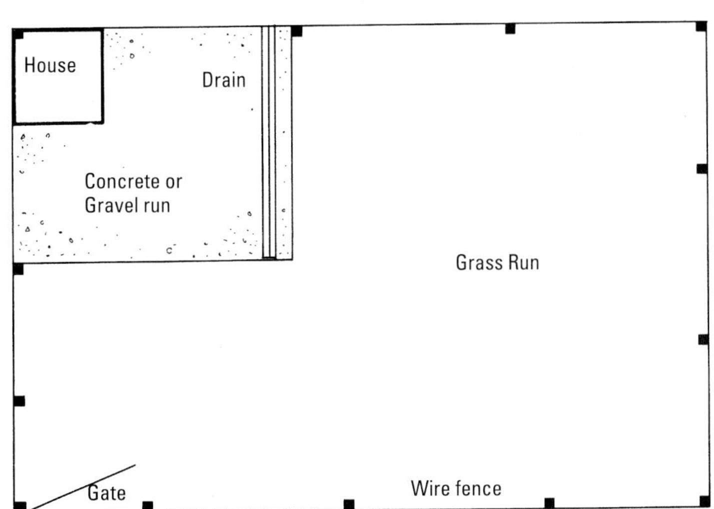

Fig 138 Plan view of a duck house with concrete or gravelled run for wet weather. This can also be made into a vermin-proof area so that the ducks can be left out overnight if necessary.

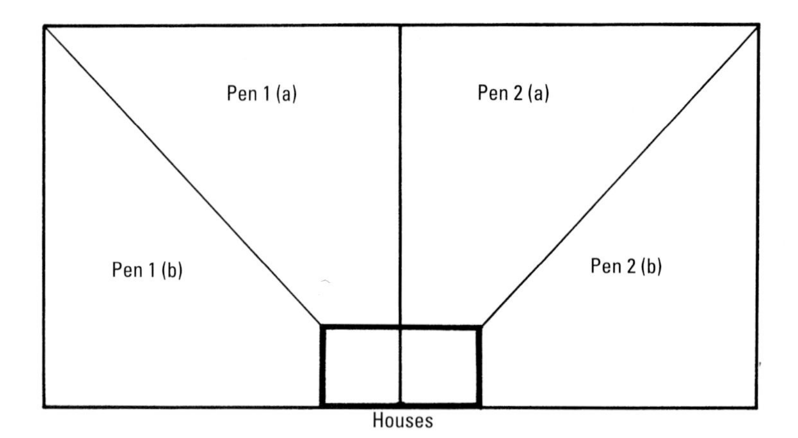

Fig 139 If more space is available, it is wise to have a dual run for the ducks. This means that only one area is in use at a time; one pen can be rested and the grass has time to grow and recuperate.

Fig 140 Strong corners are essential for the fence to last. Use braced 'Yorkshire corners' or extra support.

Fig 141 The outward overhang is designed to stop climbers and foxes from running up the wire. This is especially useful if there is no electric protection.

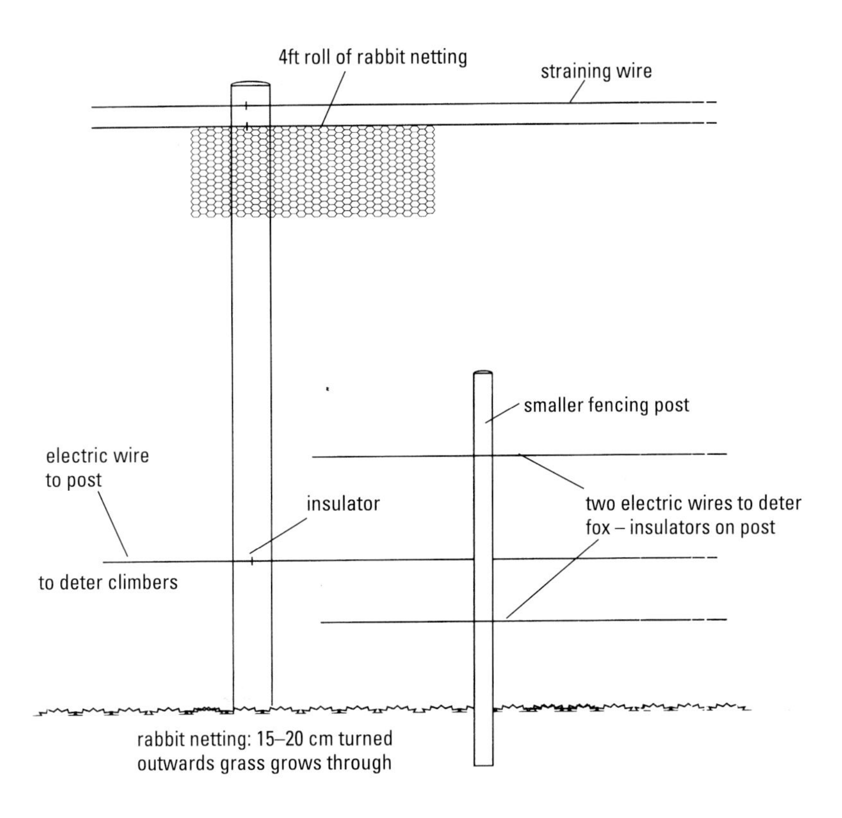

Fig 142 A lower fence is easier to erect but must be protected by more electric wire. Rabbit wire is turned out at the base. It can be anchored by turf or the grass allowed to grow through it. Two or three strands of electric wire on posts set back 11in (30cm) stop foxes trying to dig or jump but there is still the possibility of an athletic fox clear-jumping it. If polecat and mink are around, a strand of electric wire will have to be placed closer to the rabbit wire.

peoples' fox-proof pen design and get estimates before making a decision.

A rabbit-wire fence made from a 4ft (1.2m) roll but with substantial electric protection too, is easier to erect. For one breeder who has this type, it has given total protection for the birds for over three years now. This fencing affords protection from diggers and climbers, but there is always the possibility of an athletic fox clear-jumping it. However, if day-time foxes are a problem it is well worth trying as a cheaper alternative to the high fence. The main disadvantage is the high dependence on electricity. Mains supply gives a more dependable 'kick' than battery. However, back-up battery supply is useful for any night-time cut-off from the mains supply.

Electric mesh netting can also be used as a deterrent to foxes but contact with ducks and geese is not recommended. Although the birds do learn to avoid the wire, there have been cases of birds getting entangled in the mesh and dying. Use this mesh on the outside of light wire netting to keep the birds off the electric net. Check the fox-proof pen each day for signs of digging by badger or fox, and check that the electric wire is clear of interference causing earthing.

Predators and Pests

Mink, Marten and Polecat

Mink are the most common predator among these smaller mustelids. Like the polecat, they have little fear of people and will face up to humans aggressively. They are common in many areas of the UK but are more in evidence near rivers with fish. Polecats and martens used to be confined to small populations in upland areas of Wales and Scotland, but are now becoming more widespread. Stoats may occasionally be a problem for ducklings and small wildfowl. Escaped ferrets can be even more of a problem.

Protect the birds by using a strand of electric wire close to fencing poles, as these mustelids will climb. The wire mesh of the fence must be fine – 1in (2.5cm) or less. Make sure that wire on sheds is secure and that the wood is sound; these mustelids can get in through very small holes.

In general, small mustelids are easy to catch. Unlike foxes, they will enter traps easily and can be caught with fishy baits or entrails, in live-traps. There is then the chance to examine the prey before deciding what to do with it.

Fox

Foxes come in two types: the urban escapee and the

country fox. Urban foxes are known to be 'dumped' in the countryside and are a particular nuisance as they do not know how to survive on wild food. Also, they are not very apprehensive of people. They are more likely to be seen in the daytime, and are also easier to catch in a baited, live trap, than country foxes. If necessary, consult local farmers and gamekeepers about dispatching them.

Badger

Badgers are becoming more common, and adequate housing and fencing (*see* above) may be necessary to prevent a nocturnal attack. Never leave eggs or a sitting duck near a boundary fence. Badgers love eggs and will break in to get them; it is best not to leave the temptation. The badger may brush under electric wire, the coat affording protection. One way to train aversion is to fix food to the electric wire to ensure the badger does get a shock, and leaves the birds alone in future.

Dogs

Dogs can be more of a problem than wild animals, depending on the area in which you live. On agricultural land in the UK, dogs must be under the control of their owner, i.e. on a leash or within a certain controlled distance. Action can be taken against unaccompanied dogs that molest or kill the stock on your land. If this happens, record the incident, e.g. with another witness or photograph the evidence. Impound the dog if possible and report the matter to the police. Action may then be taken over the dog and an insurance claim made for the damage. Dogs that have killed stock once will come back.

Cats

Cats are not a problem with adult ducks except, possibly, Calls. Cats are easy to catch in live traps so pet animals can be identified. Domestic cats can usually be scared off and trained. Feral cats are a nuisance and a decision may have to be made over controlling them.

Hawks

Hawks are not usually a problem for adult ducks, though buzzards have been accused of taking Calls. In general, Buzzards are more interested in carrion and if sufficient dead rabbits and pheasants are about, then they will ignore the ducks.

Ducklings are more at risk from Sparrow Hawks but, generally, hawks take birds such as Collared Dove, which are on the wing. If the Call Ducks do not fly, they are generally safe. Small wildfowl, such as Teal, have been taken, however, even on the ground.

Other Predators

A number of other predators, including crows, jays, magpies, owls and rats will take eggs and ducklings. The Larsen trap, designed to catch magpies by using a live bird in it to attract others, also works quite well when just baited with Call Duck eggs. Also scatter broken eggs around the trap to attract them. This method spares a magpie being confined in poor conditions as a Judas Bird.

Rats can be kept at bay with trained cats and dogs but if you do not have these hunters, poisoning rats is the most effective method. Bait that is whole grain and coated with poison, is palatable to the rats. However, cut grain that is impregnated with poison is more effective. It is essential that all bird food is picked up at night so that a high rat poison dose is delivered. The poison must also be inaccessible to other animals and birds, and kept dry.

The odd rat can also be caught in sprung Fen traps, but these are dangerous for other animals unless carefully placed. Tunnel traps designed to catch mink, but placed along a wall where rats run, can catch them even if unbaited. Be careful that the trap is not accessible to ducks. The sprung ends could snap down and asphyxiate them, and Call Ducks are small enough to get inside them.

Identifying Predators

Attacks by polecat and other small mustelids can result in many birds being mauled about the back of the neck. This leaves the feathers matted and damp. Sometimes only the guts of the birds are eaten, e.g. by mink. Tooth marks may be found in the skull and neck of the birds.

Fig 143 Tunnel trap. The rat can run through but its weight on the treadle causes the sprung ends to close. The rat will have to be dispatched in this live trap. Do not get bitten when moving the trap. Handle materials with rubber gloves to guard against disease transmission. Rats can carry Weil's disease. This trap is also useful for catching mink with a fishy bait.

Foxes too will kill many birds, especially if they get into a building or a 'fox-proof' pen; the temptation of so many meals is too great. Foxes have to work hard to catch their food and so, at a time of plenty, will make food larders. The prey is buried and dug up when required; rotten food is no deterrent to a fox. Out in the open, however, only one bird may disappear without trace.

Dogs will hunt birds down and grab them by the back of the neck and back. They are less likely than foxes to carry the bodies off, but both dogs and foxes can leave many casualties. Dogs, judging by the misfortunes of breeders with years of experience, are more likely to leave many injured rather than dead birds compared with a fox, which is more practised at dispatching prey.

If there is a patch of many feathers in the grass but nothing else, this indicates a hawk has hit the bird. If there are dead birds, try to assess the cause as this may help you in deciding on a course of action. Note that vermin control is a specialized topic in itself; get a book on the topic (e.g. Roberts, 1989) and ask advice on rats from the local council pest control department.

Severe Conditions

Heat Wave

Ducks cope with hot weather better than one might at first imagine, as long as common sense is used. Make sure plenty of water is available if the ducks do not have free access to ponds. Shade must be available at all times of the day as the sun moves round. The best shade is provided by vegetation that the birds can sit under at any hour. Adult ducks are better at standing up to heat than ducklings, which more readily suffer from sun-stroke, and also enteritis, which is more common in heat waves. Birds should also be protected from stagnant water in hot weather as they are at risk from botulism and bacterial infections.

During prolonged hot spells, when night temperatures are also high, it is particularly important to shut birds up late and let them out early, especially if the sun is shining on the shed and it is small. Check that ventilation is adequate in these conditions too.

It is now generally accepted that the global climate is warming, whether it be the greenhouse effect, solar activity or the two combined. It would therefore be prudent to adapt the ducks' environment to cope with hotter, drier summers, especially in the South East of England, by planting more trees and shrubs for shelter. At the same time, the climate in the north and west may also be getting wetter in the winter, with consequent implications on saving the ground from damage by the ducks in wet spells.

Freezing Weather

Well-fed ducks cope better with average British winter conditions than summer heat-waves. During a rare spell of freezing weather, when daytime temperatures are near or below zero, the drinking water situation has to be monitored, as with any livestock. Empty all buckets of water in the evening and start with an ice-free bucket in the morning. Abandon feeding wheat under water in an icy spell; feed dry wheat and pellets mixed together in a bowl. Make sure the ducks are well fed; shutting them up at night in a warm shed with plenty of dry bedding reduces their food demand.

The birds can continue to live out in a fox-proof pen because they avoid frostbite by tucking their feet up into their down. Standing water may be dangerous. It can freeze over and freeze the ducks in; this will cause frostbite and the affected area of the leg or web may become necrotic. Ducks may also dive in a central deeper area that is ice-free and emerge under ice at the edges. Keep the birds away from this hazard if possible.

Maintaining Ground

When ducks bare the ground, a weed problem can develop. Try to maintain turf by sowing grass seed on bare patches in the spring and September. A good covering of grass promotes infiltration of the rain, and reduces the effect of the ducks sealing the surface by puddling clay. A thin top-dressing of calcified seaweed will maintain a neutral pH 7 (best for grass growth and worms) and give the ducks trace elements, beneficial for their health. Areas of higher rainfall in the west of Britain suffer from more acid soils and this may affect the calcium intake of the ducks.

Duck manure tends to make the grass grow rather dark green and coarse and it can get too long if it is not grazed by geese or sheep. It will need cutting to keep it fine and short, and to discourage weeds.

If plantain, daisy and nettle begin to appear, it is best to spray these with a selective weed killer that leaves the grass unharmed. As the instructions with weed killers recommend, remove the stock from the area for about a week. Spot-spraying nettles with stock in the pen can be done if essential, since the ducks will probably not try to eat these. No adverse effects have been noted from stock in close association with such weed killers as long as ragwort, which is poisonous, is not present.

Part 3
Breeding Ducks

6 Selecting and Managing Breeders

The constraints on producing pure breeds are quite different from producing commercial layers and table birds. This chapter will focus on producing good quality, healthy pure breeds. In this case, all of the eggs laid will not be set. There are two reasons for this. First, it is best to allow a two to three week gap between settings so, as the ducklings vacate the rearing quarters for a run, the next batch of hatchlings can move in. This minimizes the amount of equipment needed, including incubator space, and makes the most effective use of it. If your interest is in quality pure breeds, there is no point in rearing huge numbers of ducklings. There is a limited market for birds that are not destined for the table.

Second, not all of the birds kept need be used for breeding. The most successful breeding programmes are those that concentrate on using only the best matings, whilst others birds (both male and female) are kept in reserve. They are reserve birds in case of infertility, or are being grown on for the following season's potential breeding or show stock.

Breed Characteristics

If the main aim is to produce pure breeds for exhibition or for sale as a pure breed, the parent stock should conform to standard. It is advisable to check the birds against the *British Waterfowl Standards* and birds in breed classes at the larger shows. Buy stock from a reputable breeder who will be able to advise you on the finer points.

Sound Stock

It is very important to use birds with no physical defects such as a twisted neck, roach back, wry tail, crooked toes and legs, crooked bill or bad eyesight; these defects will breed on. Defects can crop up in any variety with too much inbreeding, but are more common in Crested Ducks (spinal defects) and Runners (crooked neck). It is therefore essential in these breeds to introduce unrelated stock if there is any sign of a physical defect. The health of the stock must come first, before the breed characteristics. There is no point in breeding perfectly coloured birds if they carry a deformity.

Age

Ducks can live up to ten years old in good conditions, and there are cases of birds living even longer. However, breeding ducks are generally between one and four years old, and drakes the same. Egg-laying capacity and egg-shell quality tends to decline with age. In heavy breeds, such as the exhibition Aylesbury, fertility in the drakes is best at one to two years old and a male may become infertile after that. A pond always helps fertility, particularly with the bigger ducks, because mating is easier.

In breeds where size is important, and in the Indian Runner where leg strength in important too, a duck in her second laying season is better than a yearling. Light ducks are considered 'mature' at about six months when they can come into lay, but continually breeding from the youngest stock, i.e. using a young duck as the breeder each year, will downgrade quality. The eggs of a young duck are often slightly smaller than the standard size, and although they do improve over the spring season, the advantage of early hatching for quality stock is then lost. For breeding vigorous birds, the advice of many duck breeders has been to use a fully mature, two-year-old female. Her eggs are slightly larger than a yearling's, and she produces stronger offspring. A young but early hatched drake has often been recommended for mating the mature duck. However, as long as an older drake is fertile (especially if the pair or trio are established), the age of the drake probably makes no difference to the quality of the off-spring.

Selective Breeding: Choosing the Breeding Pair

Selective breeding means that the gene pool of a group is restricted, so that relatives that exhibit desirable traits are used in a breeding programme. Pure breeds are thus produced by human selection of the breeding combination. This has resulted in the production of specific breeds of cattle, pigs and sheep that are suited to differing physical and economic circumstances. A similar process has also taken place with ducks over hundreds of years in the case of the Indian Runner, and over a shorter period of time with breeds such as the Khaki Campbell and Orpington.

Because purposeful selection is involved, it is useful for the breeder to understand the principles behind recessive and dominant genes (especially for the inheritance of colour), and Mendel's laws of inheritance. Details are beyond the scope of this book, but the principles are covered in GCSE Biology courses today.

The purpose of selective breeding is to improve stock, be it for aesthetic or practical purposes. Thus there is no point in using inferior birds. The foundation stock must be healthy and exhibit some particular characteristic that the breeder wishes to enhance. Thus if in a breeding pair the characteristics of the drake are particularly desired, then he would be mated back to his best daughter who also showed these characteristics. It would be similarly so with a mother and son.

Selective breeding therefore often involves a certain amount of inbreeding, i.e. using related parent stock to retain a phenotype. The danger of using related parents is that 'inbreeding increases the proportion of progeny having homozygous recessive alleles, and therefore the expression of the recessive lethal genes' (Stevens, 1991). So, as well as retaining desired characteristics, the progeny of an inbreeding programme might also acquire unhealthy characteristics. Inbreeding programmes are therefore best pursued where the foundation stock is unrelated in the first place and produce a first generation of offspring who are all healthy.

Inbreeding in the hands of breeders who have kept the same strain for many years can produce birds that breed well because the genotype of the birds is known through experience. A new breeder does not have this background of experience, so a few simple guide lines should be followed.

- Birds showing any physical defects should never be used.
- Preferably the whole hatch should be fairly trouble free. If ducklings have any deformities or growth rate problems, then those siblings which appear to be healthy can carry that defect to be passed on in the next generation.
- The health and vigour of the strain should be as important as the show characteristics.
- Poultry breeders of years of experience consider that no more than four generations of closely related breeding should be used before reverting to out-crossing to another individual of a different strain.

Behaviour in the Breeding Season

Most breeds of ducks only lay a few eggs in the autumn. The Campbells are an exception because they can lay over 300 eggs a year. In the other breeds, the ducks are most prolific between February and July. The length of this laying season and the number of eggs will depend upon several variables such as the age of the duck, the breed and strain, the weather and also the food supply.

Once the ducks come into lay, they will spend more time courting the drakes. If the drakes have not been separated for the winter season and several

Fig 144 Ducks often lose their head feathers in the spring because of the drakes holding them in their beak in order to mate. Check these birds to see that they are not injured. Heavy ducks can also lose feathers on their back from the drake's claws scratching them when 'treading'. In cold wet weather, the duck must be removed from the drake, and the feathers allowed to grow again on the back.

have been left with the ducks, then surplus drakes must be removed from the duck area. Ducks that are in lay will be repeatedly mated by drakes. They end up with muddy backs, feathers pulled off their head and neck and, in the worst cases, have their eyes damaged and may even be killed. This is more likely to happen with the light breeds of ducks and Indian Runners. Indian Runners are the most delicate breed; the ducks will go 'off their legs' and die if not rescued. Avoid this situation even developing; consider that in managing all farm animals the ratio of males to females always has to be monitored for the health of the females because, unlike wild animals and birds, they are in close association with each other all the time.

Separating Breeding Birds

Apart from the Muscovy, all of the domestic ducks are derived from the Mallard. They can all therefore interbreed. The ducks and drakes have to be separated into their breed pens in the spring to produce pure breeds. However, some of the birds are so different in size that it is possible to fence two breeding groups into the same area. Call Ducks can be kept in the same breeding pen as Aylesbury or Rouen, for example. They can generally be kept with Runners, though a small Trout Runner can mate with a large Call. If the sizes are similar, e.g. Appleyard Miniature and Call Ducks, there will be cross-breeds, however well established the pairs.

In general, large drakes do not harass Call Ducks; the Call females can even remain unmolested if they get into a pen of large drakes in the winter. However, there are exceptions to this rule. We have had Runners brought up only with Indian Runners who fail to tolerate any other breed. The background of the birds therefore determines their behaviour. A more likely case however, is that the Calls will chase off large drakes by the tail.

Breeding Groups: Numbers of Ducks and Drakes

Pure breeds are not usually required to breed in large numbers and are often kept in pairs. Drakes are usually well behaved in this situation because they are not competing with another male. If, however, the duck is being damaged by feathers being pulled from her neck and head in excessive amounts, put another couple of ducks into the same pen. Any other ducks of similar size will do, as long as you can recognize the eggs intended for breeding.

When larger numbers of fertile eggs are required, the ratio of ducks to drakes will depend upon the time of year, the breed, and the age and condition of the drakes. If early fertile eggs are required, in January and February, then a larger ratio of drakes to ducks may be needed. Fertility can be poor when the weather is cold and many eggs can be wasted as 'incubator clears'. It is often better to use or sell the eggs for eating at this stage.

A suitable ratio for ducks and drakes is three to one for the larger breeds such as Rouen and Aylesbury. Pekins are more active and can cope with up to four ducks, whereas in the light breeds one drake may be sufficient for six or even more ducks. However, there is no set rule; fertility of the eggs on incubation is the only sure guide. Heavy drakes sometimes need to be persuaded to be a bit more

Fig 145 and 146 Cross-breeds will result if ducks and drakes of similar sizes are not separated in the breeding season. This Indian Runner drake is mating with ('treading') a Rouen duck (left). After copulation, the duck washes vigorously, and the drake swims rapidly around the pool (right).

active and it may be advisable to keep two drakes with about five ducks. Males then compete and mate more frequently. Young drakes tend to mate more often than old ones but this can be a nuisance if it results in injury to the females. Also, a very active drake is no guarantee of success; he can turn out to be completely infertile.

Unlike geese, it is possible to put most ducks and drakes together and they will mate and breed. However, there are exceptions. Birds that have been paired for a long time will try to get back to each other and want nothing to do with a new mate. Calls in particular are very choosy about their partner and they must be paired up in the late autumn for a successful early breeding season.

Unlike the other breeds of ducks, Calls are usually better in pairs or, occasionally, in well-bonded trios, where females have been together for a long time. Nevertheless, Call drakes will still have a go at mating with another drake's female, so colour crosses need to be avoided by separating them into colour breeding-pens in spring. Occasionally a Call female may have to be rescued from a particularly active drake but in general the females are good at giving short shrift to unwelcome advances. In the autumn, females will even protect their own drakes during disagreements in the flock.

Nutrition in the Breeding Season: the Importance of Greens and Protein

Commercial ducks will lay all year round if kept in the correct conditions. Most pure breeds, such as exhibition Runners and Pekins, lay between February and July. Exhibition ducks and those selected for breeding should not be pushed into lay in the autumn by feeding a high-protein diet. They need time to grow instead. That is why they should be kept on maintenance pellets and wheat.

Breeding ducks need a different diet to the maintenance pellets, designed for when ducks are not in lay. This breeder diet also applies to the drakes. Both sexes need a diet richer in protein, vitamins and minerals.

Breeder pellets will produce better results than ordinary layers' pellets. Although the layers' pellets will provide the correct level of calcium for laying ducks, they do not have the correct proportion of nutrients to produce a healthy embryo from the egg of the duck or sperm of the drake. In chickens, beak and leg deformities result from dietary deficiency of vitamin D, calcium and phosphorus; curly toe paralysis develops from lack of B2 and lack of vitamin E results in brain disorders in chicks. Whilst there seems to be no documented evidence for waterfowl, it is likely that dietary deficiencies would also affect them adversely.

Some breeders undoubtedly do get good results from using only hen layers' pellets. One top breeder has used nothing else but layers' pellets for his ducks, for all stages of growth, throughout the year. However, his birds have always had free-range on top class pasture and access to wheat. The birds have therefore been able to find their own protein supplement from worms and insects, have had grass for minerals and vitamins, and could select the amount of wheat they required. When birds are kept in confined conditions, one has to be more careful about their restricted diet and a balanced ration is essential for good results.

However good the manufactured food, there is no substitute for fresh food for ducks. Fertility is

Table to compare a manufacturer's breeder ration with a maintenance ration

	Duck Breeder	Maintenance
Oil	3.75 per cent	4.5 per cent
Protein	17.0 per cent	14.0 per cent
Fibre	5.5 per cent	7.5 per cent
Ash	11.5 per cent	8.5 per cent
Methionine	0.35 per cent	0.29 per cent
Moisture	13.8 per cent	13.8 per cent
Vitamin A retinol	13,500iu/kg	9,000iu/kg
Vitamin D3	3,000iu/kg	2,000iu/kg
Vitamin E	80iu/kg	32iu/kg
Sodium selenite (selenium)	0.31mg/kg	0.24mg/kg
Copper sulphate (copper)	25mg/kg	20mg/kg
Main ingredients in descending order	Wheat, soya bean extract, wheatfeed, barley, soya	Wheat, wheatfeed, sunflower extract, maize gluten feed, barley, rape seed extract
Ingredients for extra vitamins and minerals	Fishmeal, di calcium phosphate, salt, methionine, sodium bicarbonate	Calcium carbonate, di calcium phosphate, vegetable oils, salt

Specifications from a different company, comparing breeder and layer pellets

	Duck Breeder	Layer
Oil	4.50 per cent	3.5 per cent
Protein	16.00 per cent	16.00 per cent
Fibre	6.50 per cent	7.00 per cent
Ash	11.00 per cent	15.00 per cent
Methionine	0.26 per cent	0.35 per cent
Moisture	13.80 per cent	13.80 per cent
Vitamin A	10,000iu/kg	8,000iu/kg
Vitamin D	2,500iu/kg	2,000iu/kg
Vitamin E	25iu/kg	10iu/kg

sometimes difficult to get in pure breeds and the chances are optimized by giving a varied diet, preferably with some animal protein, because ducks are omnivores. Fishmeal is now often absent from pelleted food but worms and greens may make up the difference. How important the diet is can been seen when a pair of jaded birds kept in poor conditions are then given free range and, within a few weeks, become totally different ducks on grass and wild protein.

Infertility

Even if the diet and conditions for the birds are right, there may still be problems with fertility. Poultry breeders seem to find that if one hen egg is fertile, then most of the eggs will be fertile. Goose eggs are completely different; there may be only one fertile egg in a large batch. Duck eggs can be like this too, but fertility is usually better than in geese and can be 100 per cent.

Conditions that lead to infertility can be:

- an inadequate diet – check that you have a breeder diet and fresh greens and protein for the ducks (*see* above);
- lazy drakes – solve this by putting another drake in the pen to provide some competition (there must be enough ducks for the two of the males);
- cold weather – this seems to put the drakes off mating successfully;
- prolapse in the drake – the penis of the drake is extended and cannot be retracted (*see* Appendix);
- old drakes – light drakes can be fertile up to seven years, but heavy breeds need youngsters;
- fat drakes – this is more of a problem in the heavy breeds such as Aylesbury and Rouen;
- too much inbreeding – it is thought that embryos fail to develop or start and fail at an early stage because of lethal genes;
- poor egg handling and incorrect storage.

Duck Problems in the Breeding Season

The majority of females have no problems at all with laying eggs throughout their life, but a few individuals can die. The main problems that arise with the females are egg-binding and prolapse of the oviduct.

Egg-Binding

If a small number of ducks are kept, it will be noticed if a particular duck is not laying. If she also looks dull, and has become inactive and has ruffled feathers, it is a sign she is ill or egg-bound. Check the vent. This area is normally clean, but if some droppings or mucous remains there, feel the lower abdomen between the pelvic bones to see if an egg is stuck low in the oviduct. An egg in this position can be felt and gently pushed outwards. Only try this when you have kept the duck in a warm place and applied olive oil as a lubricant on and in the vent (*see* Appendix). This is only of use if the egg is low down in the oviduct.

Frequently an egg is stuck higher in the oviduct and cannot be felt. If in doubt, consult the vet who might also be able to give extra calcium (necessary for muscle contractions) in the form of an injection. Ducks can be saved by prompt action. It is preferable to get the egg out whole to avoid damage to the lining of the oviduct. Always follow up with a course of antibiotics if damage is suspected. If the problem is high up in the oviduct you may not be able to save the duck and she is best put down.

Prolapse

Sometimes ducks strain to pass their egg, and this results in the lining of the lower oviduct being forced outwards as well. Often this is not noticed until some hours later when the flesh has started to dry and become dirty. One can try to treat the duck by washing the affected part gently in luke-warm water and trying to replace the affected area. However, it is often necessary to put down the duck before a bad infection sets in.

Prolapse of this kind seems to be worse in wet, cold spring weather when the birds are in full lay. It might be better in these conditions to avoid trouble by under-feeding pellets to try to cut back on egg production, though this frequently makes no difference. Do not use pellets with a high proportion of peas and beans; this is thought to encourage prolapse in ducks. If you have a shed, shut the birds up on comfortable dry bedding at night to keep the in-lay ducks warm. Also, remove any surplus drakes.

Thin-Shelled Eggs

With a good pellet diet, free range and access to poultry grit as well, fit ducks should lay eggs with a good shell. Lime the ground if several ducks are affected. Occasionally a duck may persist in laying thin-shelled eggs for no obvious reason. This is most likely to happen with Call Ducks. Although it may be possible to improve egg structure with calcium supplements such as Calcivet, this is probably not a good breeding strategy because offspring may follow the same pattern.

No Eggs

Occasionally a young duck is just not a layer. Sometimes a duck may lay for one year or two, then produce eggs the size of marbles. Again this is most likely to happen with exhibition Calls that are near the limits of viability. Problems with egg quality and egg number tend to come from inbreeding. This is probably the case with females that change sex after a year or two, and stop laying.

7 Eggs and Incubation

Ensuring Pure Breeds

Fertility and hatching are easy to get in cross-breeds because they are unrelated. If the ducks and drakes have been running around in a mixed flock, fertility will probably be excellent, but there will be few pure-bred ducklings. Once the ducks have been shut in their breeding pen, i.e. in a pure-bred pair or trio, then up to eighteen days will be needed to ensure that the eggs produce the desired breed. Cross-breed embryos are much more vigorous than pure-breed ones.

If a drake is taken away from a pure-breed pen, then the duck eggs will probably remain fertile for up to fourteen days after the drake has been removed. Check this for yourself by incubating dated eggs from such a breeding-pen. It is useful to be sure of the times for separating the birds because the same duck can be used with a different line of drake in the same breeding season. This can produce two half-related lines of offspring from which to choose the next year's breeding birds. Not all breeds are willing to accept this. It will not work with Call Ducks, who will not accept a new mate easily.

The Best Time of Year to Set Eggs for Hatching

Quality

If quantity rather than quality is important, then there is no point in selecting particular times for hatching. However, in Runners and heavy ducks, the strongest ducklings and best growers are usually hatched from the earlier eggs laid by the mature ducks. This can mean eggs set in April for mature ducks, which lay later as they get older. Even with light ducks, late-hatched birds tend to be smaller than their early-hatched siblings. For some reason, the earlier eggs in the season's batch are the best. There are drawbacks to this, however:

- early ducklings often have to be kept indoors for longer because it is too cold or wet to rear them in grass runs (where they will be stronger than in intensive, heated conditions);
- also, the main problem with early birds is the frequent preponderance of drakes;
- with respect to Calls, where oversize is a problem in early hatched birds, it is often best to wait to set eggs until April – this is especially the case if fertility is low.

Ratio of Males to Females

The sex of the offspring is determined by the female, not the male, in birds. However, to try to avoid the situation of large quantities of unwanted drakes, a series of management strategies can be adopted. These environmental factors also seem to have a bearing on the male-to-female ratio of the offspring.

A mid-season batch of eggs should be set to try to get more females. A very late season batch in late May or June is often unsuccessful; many breeders find that there is again a majority of males.

In general, enthusiastic young drakes are more prone to producing males than females. This situation is especially bad if two drakes are competing over one female, whose eggs can yield a complete hatch of males. Try to avoid this situation. Keep a drake with several females whose eggs are not actually needed. Use only the eggs from the specific female wanted for selective breeding. Her eggs might be recognized by colour, or she can be shut up separately from the other females at night until she has laid (i.e. trap-nested).

Nests

If the ducks are shut up at night it is easier to collect clean eggs if there is enough room in the shed to provide the ducks with a boxed-off corner at floor level in which to lay. Encourage the ducks to lay in the nest by leaving a pot egg or a marked egg there. Keep the nest topped up with fresh shavings each day and the ducks will leave it clean too. If there are few eggs, check that the ducks are not burying them in the shed litter; they are following their natural instinct to hide their eggs.

In a fox-proof pen too, ducks like to choose a regular place to lay on the ground. It is worthwhile making attractive places for them. Provide nest-boxes or greenery for them to hide under. A pile of branches screened with cypress for shade and cover is appreciated. Failing that, make a tent by tying an old sheet over a pile of branches. The ducks like to go underneath this, and it discourages them from dropping their eggs in the open. If cover is not provided, jays, magpies, jackdaws and crows may all reach the eggs before they can be collected.

Nest Materials

Nesting materials must be fresh to keep the hatching egg as clean as possible. White wood shavings are good because they do not harbour mould. Outdoors, in a wet spell, something more robust, such as wood peelings from larch, works even better. Straw tends to absorb too much moisture and go mouldy. Put plenty of litter down in the favourite spot to try to keep the egg off the earth. If the protective mucous of the cuticle can be preserved intact, without any contamination by moulds and droppings, this is the best type of egg for hatching.

Collecting Eggs for Incubation

Always try to collect the eggs between about 7.00 and 9.00am because most are laid at this time by the larger ducks, which generally do not go broody if they are not encouraged to do so. If there are fewer eggs than expected, check that they are not buried in the litter (*see* above) and that the ducks are not being let out too early. If they are released just as they are about to lay, the eggs are often dumped in the pool.

Calls and Appleyard Miniatures are different from the larger ducks in their laying habits. These smaller breeds can lay at any time of day and need encouragement to keep laying in the same nest. This is especially important if these small ducks are to go broody. If all of the eggs are removed, they will become disgruntled and dump them on the grass, or use another place that is more difficult to find. Collect the valuable eggs and leave marked, substitute eggs (which are not needed) in the nest to encourage the bird to continue laying in that place. Unless the birds are kept in netted aviaries, well-camouflaged nests and regular egg collection are essential to beat the magpies. Pest control may be needed too.

Carry a suitable container for egg collection such as a bucket part-filled with shavings. This stops the eggs rolling around; they must not be cracked or dented and should be handled gently. Make sure that your hands are clean for handling the eggs too, so that bacteria from dirtier places are not transferred to clean eggs.

Label the eggs as they are collected with the date and name of the breed or duck-pen. Use a soft 2B pencil because hard pencils do not mark eggs that have a good covering of dried mucous very easily. Spirit markers can be used for a more permanent label if the eggs are destined for natural hatching. Mark these on the bulbous air sac end of the egg to keep chemicals a bit further away from the developing embryo.

Cleaning and Disinfecting

Clean, dry, unwashed eggs are much the best for hatching because the dried mucous from the oviduct, which covers the egg, provides a barrier to bacteria. If the eggs are destined for natural hatching, they are far better left unwashed. The protective layer is removed by washing and makes the eggs more porous. Loss of water at a greater rate from the egg is not, however, a problem in the UK, and slight loss may even be advantageous.

If the weather is wet, it is difficult to collect clean eggs. Some will be muddy if laid outdoors; some will have droppings on them if laid in a shed. If there is only a small amount of contamination, wipe off dirt with a clean, dry cloth or tissue if possible. Muddy eggs can often be cleaned gently with a cloth, or the mud can be rinsed off under a lukewarm running tap. The natural mucous is not all removed in this way.

Dirtier eggs should be thoroughly cleaned straight away, but reject any eggs that are very dirty – they will almost certainly become a health hazard in the incubator. Moulds and bacteria such as *Escherichia coli* are picked up from the litter of shed floors, and enter the pores of the eggs more easily under damp conditions.

Wash the eggs in lukewarm water, which is warmer than the egg. This is so that the shell membrane does not contract and draw bacteria into the pores. Eggs laid out in the rain and in puddles are of no use because of this. Scrub dirtier eggs with a nail brush under a running tap, or use water with a drop of Milton. Dry the eggs quickly and thoroughly by draining them on a wire rack or wiping with clean tissue. Virkon is also recommended as a safe bactericide and will help prevent dirty eggs going rotten and contaminating the incubator.

When washing the eggs, make sure that they are not rotated in the same direction as they are cleaned. This winds up the suspensory structures (the chalazae) inside the egg and may ruin it for hatching.

The warm incubator provides ideal conditions for

bacteria on dirty eggs to multiply and so formaldehyde gas, generated from potassium permanganate and formalin, is used to fumigate incubators and hatching eggs in commercial hatcheries. These materials should not be used in the home, only in an outdoor well-ventilated shed. The quantities and containers used (they should not be metal) should be checked with the supplier before use. For small quantities of eggs, the care outlined above is generally sufficient to achieve a good hatch without resorting to formaldehyde. Washing the incubator with Virkon (avoiding the metal parts) is also a safe and effective alternative.

Selection of Eggs for Incubation

- Only store those eggs for hatching that have the best chance of producing top class birds. If you have young ducks that are laying small eggs, there is no point in using these if you want quality birds. Wait until a bit later in the season to see if egg size improves; use mature birds' eggs if possible.
- Choose eggs that are the average size for a mature bird of the breed. Reject over-sized eggs, which are probably double-yolkers.
- Also reject eggs with irregular or thin shells, or eggs that are an odd shape.
- Cracked eggs are no good either – they admit bacteria and dehydrate rapidly. Even mending valuable eggs with nail varnish as a sealant rarely works.

Storage

- The eggs should be stored at a steady temperature between 12 and 14°C. Temperatures over 21°C will allow slow, weak development of the embryo, which is then rendered useless.
- Stored eggs begin to lose water through their pores. Excessive moisture loss will cause growth of the air sac and also cause problems for the embryo on incubation. Humidity should be kept about 75 per cent RH, which is usually obtained in Britain in an unheated north-facing room.
- Store the eggs vertically (air sac uppermost) and propped up against a wall to save space. A box of sand or a new, clean egg tray can also be used for this. The eggs need not be turned for seven days if stored in this position. If you wish to turn them, an egg tray saves time; it can be propped up on a slant in one direction one day, and in the opposite direction the next.
- If the eggs lie on their side, they must be turned once a day (an opposite direction each turn). This is to prevent the yolk (which is lighter than the albumen) and the germinal disc (which always floats upwards, *see* page 150) from sticking to the shell's membrane.
- The eggs should be less than ten days old for optimum results in the incubator, and seven days is preferable. Fourteen-day-old eggs may start in the incubator, but the embryo is more likely to have failed after a week. Eggs that have just been laid have not lost any water in storage and, in a hatch where insufficient water loss is a problem, these can be the least likely to hatch. It is best to let an egg rest for a day after laying before it is put in the incubator.
- Eggs up to twenty days old can be used for natural hatching. Call Ducks frequently decide to go broody after a clutch of six to twelve eggs, but the eggs may have been laid over more than twelve days. Calls are not as regular at laying as Runners.
- If eggs have been transported, allow them to settle for twenty-four hours prior to incubation. This seems to help hatchability after sending them through parcel post, for example. Posted eggs can hatch but they have sometimes been shaken so severely that disruption of the air sac membrane can be seen on candling. These eggs will not start in the incubator.

Natural Incubation

Ducks

The Best Sitters

Ducks are generally poor sitters and mothers compared with geese and hens. Indian Runners and

Fig 147 Store the eggs vertically or in a tray so that the whole batch can be tipped in one direction once a day and in the opposite direction the next.

Khaki Campbells have been bred for egg-laying and should not be expected to go broody. However, with a well-concealed nest in a fox-proof pen, both of these breeds will occasionally sit, particularly late in the season.

If you want ducks to sit – and they do achieve good results – the best types are a tame Mallard, Appleyard Miniatures and Calls. Nevertheless, they should be checked twice a day to ensure that they are still sitting. Eggs that have been deserted and gone cold for twelve hours will probably not die if put into a warm incubator.

Black East Indians are also good sitters but are often flighty to handle, which can lead to damage of the eggs or, later on, the ducklings. For bigger eggs, the Muscovy, which is unrelated to all the other domestic ducks, is a good, reliable choice.

The Nest

Ducks are far more likely to go broody in a nest of their own choosing and, like geese, cannot be moved from their preferred site. The nest is just a scrape in the soil, often under a hedge, and the eggs are camouflaged with leaves whilst the clutch is laid. If the leaves are sharp, like holly, or there are twigs of hawthorn or blackthorn, remove these or they will puncture the eggs. Provide suitable fresh litter such as leaves or shavings. When the duck is about to go broody, she will spend more time on the nest and begin to pluck down from her breast and underparts to line the nest.

Sitting

Laying ducks will sometimes spend several days deciding whether the time is right to go broody. They appear to mess about, on and off the nest, partly warming their eggs and then leaving them. At this stage, this does not seem to matter. Eggs are warmed and turned in the nest every day as a part of natural brooding and it is believed that this is a beneficial aspect of natural hatching.

Fig 148 This Appleyard Miniature female has chosen a good place to sit inside a run where the nest is protected from magpies. She is also in a fox-proof pen.

The preferred clutch for a Mallard or Call is six to twelve eggs. Miniature Appleyards lay far more and the egg number needs to be reduced to a manageable eight or ten. Hatchability from such nests is often 100 per cent. Despite the eggs resting on the ground and getting trodden on by dirty feet, they look perfectly clean. They acquire a good gloss of grease from the oil on the duck's feathers and skin. This affects the permeability of the shell and stops deterioration of the egg, as does the presence of a natural antibiotic secreted by the skin (Anderson Brown, 1979). In the majority of cases, the duck gets the water content of the egg correct too, though on a dry site in May or June the eggs can dehydrate too much despite the duck bathing each day.

A disadvantage of the duck sitting in her chosen nest in a fox-proof pen is loss of eggs through predation. One has to balance the enhanced hatchability of the egg against loss by rat, weasel or magpie. Also, as the eggs come up to hatching, the risk of predation increases. The hatchlings start to squeak and this may attract predators. It is essential to move the duck with her hatch into a coop and run, where they are all protected from rats and magpies.

Care of the Duck and Ducklings

Whilst she is sitting, leave the duck alone except for getting her off the nest once a day to feed and bathe, if she is fairly tame. A tame duck is a big advantage when you have to move her and the ducklings too, as she will not flap and injure them when she is handled. Drakes should be kept away from the ducklings. The larger drakes will harass and kill them, though we have had a Silver Bantam who was an excellent father. Food and receptacles for the ducklings should be as for the ducklings reared on their own.

Broody Hens

Broodies are easier to manage than ducks. A broody who is accustomed to sitting will tolerate far more interference from people than a duck, but will need more looking after. I prefer to use broody hens for sitting rather than the ducks. They can be persuaded to sit earlier, often shortly after the ducks have began to lay. Any type can be used as long as she has the broody trait. We have used Silkies (but birds with unfeathered legs are better), Silkie × Indian Game, and broody-type bantams. The smaller birds are best for duck eggs, especially Call Duck eggs, which need the lightest bantams.

The Nest

Hens can be encouraged to use suitable nest boxes in sheds, which can be shut up safely from vermin. The nest boxes must be of a 'walk in' type. Broodies can jump down onto most goose eggs without harming them, but the duck eggs are delicate and break with this kind of treatment. The nest boxes can just be partitions on a concrete or brick floor, and this location will generally give about the correct humidity for the duck eggs in the earlier part of the season. Nest material can be chopped straw topped up with plenty of shavings.

The hens are best started off on their own eggs then, when thoroughly broody, the duck eggs are substituted. This should preferably done in the evening because, this late in the day, a hen is less likely to decide to consider the nest unsuitable and leave. Do not expect the broody hen to waste her energy warming up cold eggs again; she may reject them. Give her warm eggs that you have started in the incubator and that you know are fertile. There is no point in wasting broodies.

During incubation, the size of the air sac and health of the eggs should still be checked by candling and any rotten eggs removed. If they are broken by the hen, the bacteria may infect the good eggs. Later in the season, from May to June, particularly if the weather is hot and dry, the duck eggs may need more moisture. This can be supplied by a light spray of water from a pump-action container, or from the nest material. The traditional way of supplying moisture is by using a piece of turf as the nest base. Moisture from the turf itself is gradually evaporated but can be replenished by introducing water at the edge and under the turf.

Although hens should sit for only twenty-one days to hatch their own eggs, they can manage twenty-six to twenty-eight days perfectly well if looked after. Many breeders use the hens for even longer periods than this. The health of the hen can be gauged by her weight, if handled daily. Her well-being can also be assessed from the colour of the comb and her behaviour on feeding.

Feeding

Hens that are used to you become very tame when handled as broodies. Lift them off the nest twice a day to feed if the weather is fine or hot. Feeding once a day will do if it is raining, as the hens do not appreciate getting wet and wind-blown. Wheat is usually recommended as a good broody ration, as the droppings are firmer and less messy should the nest be fouled. Birds do like wheat that has been soaked in water, so leave some in a bucket overnight ready for the morning feed. It slips down more easily than dry rations and ensures that the hens do not get dehydrated in heat waves. Water should always be made available, nevertheless. If household scraps are often used to feed the hens, these should be limited if they contain salt as this will place more stress on the hen during sitting.

Parasites

During the broody period a hen will become infested with northern mites, which she normally keeps to a minimum with preening and dust-bathing. Whilst sitting, this is limited to the ten to twenty minutes a day spent off the nest. So, she will need help with a good dusting powder, preferably an organic, non-toxic type such as *Barrier*. It also helps if the nest litter is changed every couple of weeks as, on close inspection, it may be found to contain blood-sucking insects. Dust the nesting container before fresh litter is used, and the hen, particularly round and at the vent where parasites accumulate for the moisture. An aerosol insecticide is best for this, to direct the spray onto the vent. Do not dust the hen just before the ducklings are due to hatch; they will receive a large dose of the insecticide which may kill them. Keep the hen clean in advance.

Hatching and Rearing

Hens can hatch and rear ducklings and make a good job of it, but it is essential to observe what is happening. Never have more than one hen sitting on a nest of eggs, especially at hatching time. They will get in each other's way and are more likely to tread on and injure the hatchlings.

As the time for hatching approaches, the hen can feel the chicks moving inside the shell, and hear them squeak, so she sits tighter and spends very little time off the nest if she is removed to feed. Most hens are good mothers but some do not take to their eggs changing shape; hens are known to kill ducklings and chicks as they hatch. So with a new hen, observe how many eggs she has in the nest, and what she does as they hatch. My neighbour uses one hen exclusively as a sitter and never as a hatcher or rearer.

If left together, the hen and ducklings must be kept in a wired run and coop to protect them from crows, magpies and rats. Suitable food and water containers must be used (*see* the section on rearing ducklings). It may be preferable to give the hen some chicken eggs that are on the point of hatching (from the incubator) and take the duck eggs away on pipping to a hatcher. It is particularly important to remove Call eggs because the ducklings are so small and vulnerable. Also, ducklings seem to be prone to getting tangled up in the hen's feathers and can be strangled. This is particularly so with Silkies, which have feathered legs, so clean-legged varieties, such as a Silkie × Sussex cross, are better if the hen is to rear the chicks.

Why Use Broodies?
If you like hens and only want to rear a few ducklings, the broodies are much cheaper to 'run' than incubators. The birds have to be fed all year, but there is no electricity bill and virtually no capital outlay. Some automated incubators are expensive for very little egg space, and the broody is equally adept. Broodies have been traditionally used to incubate 'hard to hatch' eggs because, with a little help, they get the turning (many times a day), temperature and humidity just right.

Call Duck eggs are far more difficult to hatch than other types of duck eggs, and success in hatching Calls is poor in an incubator that does not have automatic turning and very good temperature control. A broody can manage both of these, but there will be some losses due to breakage. Eggs can get punctured by a claw, or crushed when the ducklings pip. Remove Call eggs as soon as they are due to hatch, i.e. they are squeaking and one or two have just pipped. It is essential to be around at hatching time or some ducklings will be crushed.

Broodies are also useful in very wet weather when, despite having run the incubator with no water, some eggs have lost insufficient moisture. The last six days under a broody may mean the difference between the eggs hatching or being 'dead in the shell' through inadequate moisture loss. To achieve this, the broody nest must, of course, be made from dry shavings, not a moist turf. Broody hens are able to achieve more moisture loss than a still air incubator in high humidity conditions in Britain.

Artificial Incubation

Types of Incubator: Still Air and Forced Air

Modern incubators run on electricity and have electronically controlled thermostats, which are excellent at keeping a steady temperature. They have the disadvantage of being out of action in a power-cut and this is why some people still prefer the older paraffin types. Curfew may be the only company still marketing this machine, which is invaluable in areas with no electricity supply.

There are two main types of electrically run machine: the still air and the forced air (also called fan assisted):

- Still-air incubators have only one 'layer' of eggs, so this limits their capacity. An element in the insulated box heats the air, which convects naturally so that it is cooler at the base of the box and the warm air escapes at the top. The temperature is adjusted so that it is correct for the layer of eggs. The box is designed to spread the warm air as evenly as possible but there is usually a temperature gradient from the sides to the middle, as well as from the bottom to the top.
- In the forced-air incubator, a fan circulates the air around the whole of the box, so there can be several layers of eggs inside a cabinet. The even flow of warm air around the eggs ensures that the whole incubator and its eggs, whatever their size, are at the same temperature throughout. These incubators are generally much more expensive than the still-air models because they are usually larger, incorporate automatic or semi-automatic turning, and sometimes humidity control. The fan can be noisy in operation; listen to the machine in operation and consider where you can accommodate it before buying one.

Methods of Turning Eggs

Eggs in an incubator must be turned automatically or manually. Manual turning must be done in opposite directions at least three times a day. Semi-automatic incubators are designed so that a whole tray of eggs can be turned by the operation of one lever by hand.

Some still-air incubators have automatic turning gear. In a side-to-side roll on a moving floor, eggs positioned between roller bars tend to march up on each other and the incubator cannot be left for long periods of time without checking up on what is happening. In forced-air cabinets, the automatic turn is usually different; the eggs are placed in trays that tip forwards and backwards. This frequent turn through 90 degrees seems to be as good at hatching waterfowl eggs as a 180-degree turn (though the larger turn may be beneficial for eggs that are difficult to hatch). The eggs themselves may be packed vertically or horizontally in the trays, depending on the model.

In addition to these two types of turn, there are now also models with a floor of rollers for a single layer of eggs plus a fan to circulate the air.

Incubator Materials

A well-insulated incubator will be cheaper to run

Fig 149 A forced-air cabinet model with four trays for the eggs and automatic turn. Some Call Duck eggs will hatch in these models – about 25 per cent compared with up to 100 per cent of Indian Runner eggs.

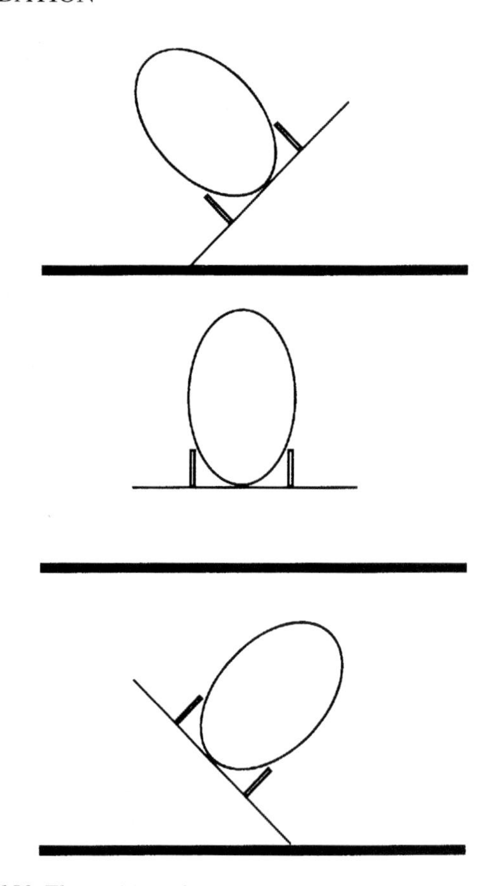

Fig 150 The position of the eggs in the trays in a 90-degree automatic turn.

than one made out of more conductive materials. Expanded polystyrene types can achieve excellent hatching results, probably because they maintain a steady temperature. The disadvantage of these is that the material is weak and does not stand up well to cleaning. In contrast, expanded polyurethane has a hard, smooth surface that is easy to clean and can be disinfected and scrubbed. Fibre-glass faced wood is not quite as well insulated as the preceding types, but will last for years.

Choosing an Incubator and a Hatcher

When choosing an incubator, there are several variables to be borne in mind such as size, cost, turning technique, materials and cleaning, running costs and positioning in the kind of accommodation you have available. If Call Ducks are to be hatched, an automatic turner with excellent temperature control is essential. Call eggs rarely hatch in small, still-air models; broody hens and ducks beat all incubators at hatching Calls.

- If incubation has not been practised before, the most useful first purchase is one that is not too expensive but will also double up as a hatcher if the hobby expands. This might be a still-air model that could accommodate thirty to fifty eggs. This may seem a large size initially if the ducks are for hobby purposes, but not all of the eggs set will be fertile. Also, hatching eggs need more space than incubating eggs.
- A second-hand still-air incubator is all right to start with, as long as it is a well-known make that is advertised in current magazines, so that spare parts can be purchased. It should be a recent model and it will need disinfecting.
- An incubator to be used as a hatcher as well must be easy to clean because hatching eggs make a mess.

Fig 151 The base of a useful still-air incubator made by Brinsea. The automatic turning gear has been removed in this model because it is used as a hatcher. The polyurethane is easy to clean and can be immersed in a bath. The eggs have been tightly packed in the incubator to start them off. At four days, when the infertiles are rejected, there will be more room. The kitchen paper makes the eggs easier to turn smoothly by hand.

- Incubators run more efficiently, at a steadier temperature and hatch eggs better if they are full. This applies particularly to still-air incubators. It is better to start with an incubator that matches the amount of eggs you wish to set in one batch, bearing in mind that eggs should be ten days old or less.
- The incubator/hatcher need not incorporate expensive turning mechanisms, if hand-turning of the eggs can be done. The turning mechanism is redundant in a hatcher anyway.

Still-Air Incubators

The section that follows is about still-air incubators for small-scale production, unless otherwise specified. Unless you are going to be content with just one incubator full of eggs that goes right through to hatching and cleaning before you start another batch of eggs, you really need two incubators for several reasons.

- If cold eggs are added to a still-air incubator, it takes some time for them to heat up. They initially reduce the temperature of the incubator and it takes hours for the temperature to stabilize again. This adversely affects any eggs already started, for example, two weeks earlier. So if you have two incubators, one of these can be used to warm up and start a new batch without affecting other incubated eggs.
- Eggs run at different temperatures: eggs near hatching generate their own heat and are considerably warmer to the touch than newly added eggs, so that in a small incubator some eggs can feel hot whilst others are not warm enough. Near to hatching, eggs are best transferred into the incubator to be used as a separate hatcher because the temperature can be regulated better by turning the thermostat control down for these hot eggs, which are 1–2°C warmer to the touch than newly set eggs. A small incubator must not overheat; 39–40°C is lethal if this temperature is maintained.
- It is inadvisable to set two batches of eggs in it with different hatching dates if no other hatcher is available. Hatching eggs make a mess and can introduce infections to incubating eggs. In a forced-air automatic-turning machine, eggs due to pip are generally moved down to the floor of the machine into the hatching tray. This introduces dirty material into the machine, so it is always better to have a separate hatcher.

The Incubation Period: Records

Throughout the incubation period, keep records of what has happened. Make a chart to fill in the date of setting the eggs, their likely hatching date, the temperature of the incubator and its relative humidity. On the same chart, the direction of the hand-turn of the eggs can be written down three times each day. Percentage fertility and the final hatch rate from the fertiles can also be monitored. If you are new to incubation techniques, a lot can be learned from trial and error.

Choosing a Suitable Place for the Incubator

There are ways of ensuring that the small incubator does as good a job as possible.

- Keep the incubator on a level surface, otherwise it affects the temperature of the eggs, which will be at different heights instead of at the optimum temperature at one level.

- A steady temperature must be maintained. This is best achieved by keeping the incubator in the house, preferably in a spare, well-insulated, unheated room. Check the temperature variation over a 24-hour period; it should be steady and certainly less than 5°C variation. A north-facing room is best, to avoid overheating through sunlight streaming through a window.
- The incubator will maintain a steadier temperature if it is covered by a folded blanket. This may impede the ventilation holes on the top of the incubator and nearer the end of incubation these must be left at least partially free as oxygen demand from the incubating eggs is higher. However, without this kind of extra insulation, it is often impossible to get a steady temperature in smaller incubators.
- This insulation also has the advantage of keeping the eggs in the dark, which is considered to be beneficial.
- The room must be ventilated; this applies especially to large incubators.
- The room must have low humidity, so that you can control the RH in the incubator. If the room humidity is above 60–65 per cent, you cannot regulate the incubator very well. Choose a room away from sources of damp air. Avoid the kitchen and the bathroom and make sure that moist air is vented from the house rather than into the incubator room.
- A cellar is successfully used by some breeders because the humidity is constant. Cool cellars and outhouses that are very well-insulated and dry, can achieve a low RH allowing much more control over egg dehydration, if necessary. Dry incubators can run at 30 per cent RH in these conditions.
- If you have young children, make sure that they do not have access to the room to change the controls.
- If the incubator room is an outhouse, make sure it is free from vermin, which carry disease. The area should not be in contact with the adult ducks because of cross-infections from the birds or their litter. Do not use the incubator room for rearing birds either.

Setting the Eggs

Timing

When setting the eggs count on to, and make a note of, the optimum hatching date. This should be a day when you are likely to be around to keep an eye on things.

Cleaning and Age of Eggs

The eggs should have been cleaned first in the manner described above. For development in an incubator they should preferably be less than ten days old.

Labelling

- All eggs should carry the date laid and indicate the parent, because any drake who is infertile can then be replaced with a reserve bird.
- The eggs will have to be turned manually in the small still-air models and so they are traditionally marked with O on one side and X around the other side. This is to check that every egg has been turned.
- If there is more than one setting in the main incubator at a time, it is wise to mark them differently, e.g. a double circle on one batch or different coloured crayon marks. If the air sac is being observed in its development, it is important to know how long the egg has been incubated.
- If you are interested in monitoring the eggs closely over incubation, record the weight of each egg in grams on the egg in pencil and on a chart.

Preparing Eggs for Incubation

This is very important.

- They should have been stored at 12–15°C.
- They should not have been transferred immediately to an environment at 37°C. A rapid rise in temperature can cause the membrane around the yolk to burst. Bring the clean eggs into a warm room overnight to raise their core temperature to about 20°C, then transfer them to the incubator.
- The incubator should have been pre-heated to 37.3°C for a couple of days to make sure that it is running properly. This is the correct temperature for the core of the duck egg and the thermometer bulb should be placed at this level, i.e. about 1.5cm above the floor of the incubator.

Temperature

For a still-air incubator, manufacturers usually recommend 38.4°C but this is for the thermometer 2in (4.5cm) above the floor of the egg tray for hen eggs during the first week, and rising to 39.5°C for the third week. Anderson Brown also recommends 38.4°C for duck eggs, the thermometer being level with the top of the egg. This will give a correct core temperature of 37.3°C. The thermometer bulb should not rest on the egg because it is the air temperature that you want to read.

It will probably be impossible to regulate the temperature in a small still-air incubator that closely. For example, if three or four thermometers are placed at different positions around the incubator (at the same height) there can be between 1–2°C difference in temperature between the sides and the middle, depending on the size of the incubator. Conditions in the nest vary too so, as long as the relative positions of the eggs are shuffled around the incubator from time to time, they will come to no harm.

Incubation temperatures for duck eggs compared with other species

	Day 1–14		*Day 15–28*	
Species	**°F**	**°C**	**°F**	**°C**
Goose	99.0	37.2	98.5	36.9
Duck	99.2	37.3	99.0	37.2
Hen	99.5	37.5	99.3	37.4
Pheasant	99.75	37.6	99.5	37.5

(Figures from Anderson Brown, 1979)

If there are two batches of eggs in the main incubator spaced at two weeks apart, the newly incubated eggs should be placed in the warmer part of the still-air incubator and the two-week eggs in the cooler part, otherwise the temperature differential between them will increase. This is because the more advanced embryos are now generating their own heat.

There is no need to worry about the eggs cooling off whilst they are turned; the core temperature of the egg lags behind the air temperature. Cooling the eggs during the mid-incubation has even been recommended. This should be once a day for ten minutes, or even up to half-an-hour, especially for forced-air incubators (Pearce, 1998).

This procedure, where the temperature of the egg is dropped to 27–35°C, has resulted in increased hatchability varying from 2 to 25 per cent, or even higher, in the case of ducks and geese. It is thought to mimic natural conditions and hence improve hatchability. Pearce wrote that it was not known if the improvement is due to changes in humidity, carbon dioxide levels or chilling alone. However, daily chilling from day eight to twenty-seven (five minutes a day increasing to ten minutes) is thought by incubator operators to help the development of the air sac, i.e. dehydration of the eggs.

The Effects of Variations in Temperature

Pearce presented a summary of H. Lundy's review of the effects of various temperature zones on the viability of hen's eggs. The outcome of this research was as follows:

1. Cooling of eggs for short periods on a regular basis at any stage of incubation, has no detrimental effect, and is probably beneficial. [Nevertheless, note point 2; abnormalities in Calls Ducks possibly occurred through too much cooling.]
2. Cooling should not be done in the early days of incubation where prolonged periods (perhaps over two hours) of 27–35°C may result in large numbers of deaths and abnormalities.
3. However, *rapid* cooling to 5–20°C may not have the same effect. Eggs incubated for one to seven days can experience these lower temperatures with little effect other than slightly delayed hatching.
4. The implication of points 1 and 2 is that, in the event of a power-cut early in incubation, it is better to cool the eggs rapidly, especially if the anticipated period is over two hours. The eggs will be in suspended animation.
5. If a power-cut occurs when eggs are near to hatching, it is preferable to limit heat loss by insulating the incubator and warming the room, if possible. The older the embryo, the more likely it is to die as a result of chilling below 25°C. Embryos that do survive, however, show no ill effects.
6. These older embryos are more tolerant of higher temperatures than newly set eggs. [This is probably because hatching eggs need to be; they generate an enormous amount of heat by their exertions on hatching.]
7. Pearce found little data specifically on waterfowl eggs, but duck eggs, and to a greater extent goose eggs, were said to benefit from periodic cooling. [It is certainly the case that eggs that seem to be very cold at any stage of development (and one might presume dead) can often be rescued by putting them into a warm incubator.]

Despite these comments on the effects of variations in temperature, it is important to achieve a steady, correct temperature at the start of incubation and to be able to maintain this. Any cooling of the eggs practised during incubation must be done by removing the eggs or opening the incubators. The thermostat controls must not be altered.

Marginally high temperatures late on produce earlier hatches, but prolonged core temperatures above 39.4°C are lethal. Mistakes with temperatures cannot be corrected by compensating the other way once a mistake has been made. If, however, the eggs have been severely heated for a short period, open the incubator and allow them to cool for fifteen minutes to get the core temperature back to the correct level as quickly as possible. Then try to keep as steady a temperature as possible for the remainder of the incubation period.

Slightly lower temperatures delay the hatch and resorbing of the yolk sac. The eggs are frequently too wet. This situation is particularly bad if humidity has been high too and there may be many 'dead in shell'. The ducklings that do manage to hatch are large and soft.

Turning

- Eggs need not be turned for the first twenty-four hours in the incubator.
- Always wash your hands before hand turning eggs, so that pathogens are not transferred to the incubator.
- After a day, turning is essential. The germinal disc (the white spot on the yolk of the un-incubated egg) is where the incubated cells will divide and grow. This disc is located on a part of the yolk that is less dense than the rest, so that when the egg is turned, the disc floats upwards. It is therefore in closest contact with the body temperature of a sitting bird. Turning the egg not only prevents the disc and yolk sticking to the outer membrane of the egg but also places the disc in contact with new nutrients. When the embryo has scarcely developed its own blood supply it needs an immediate nutrient source from adjacent fluids.
- Eggs are turned frequently by broody birds, once recorded as every thirty-five minutes on average. In automatic incubators they may be turned once an hour.
- Turning twice a day by hand is an absolute minimum and the turn should be through 180 degrees; three or five times a day is more effective. The odd number is useful because if the time intervals are not equal then they will be compensated for on the next day.
- Turning the eggs alternately left then right is essential. It is important not to keep winding the egg in the same direction as the chalazae (the rubbery strings supporting the yolk in the centre of the egg) will gradually get tightened up at one end, and get too loose at the other. (This should also have been taken into account when washing the eggs.) Insufficient turning of the incubating egg will result in the embryo dying. However, turning is unnecessary in the two days before hatching.
- It is best to get into a routine for turning the eggs, so that they do not get forgotten. Also, always write down what has been done. List the dates, indicating the date of incubation and the likely date twenty-eight days later for pipping. Opposite each date, write down the left/right /left turns each day, then there is no mistake over if the eggs have been turned or not.

It may be a nuisance turning the eggs, but it is a very good check on how things are going. Operators get accustomed to the correct feel of the eggs and can even gauge, like the sitting hen, which eggs have died and which are healthy by their temperature in the hand. This is because, as incubation progresses, the embryo begins to generate its own heat. By fourteen days the eggs feel a bit warmer than those incubated for one to six days. After two weeks, the egg's own metabolism raises its own temperature above that of the air in the incubator.

Humidity

Measuring Humidity

Air contains water vapour as a gas. The amount the air will hold depends on its temperature. The easiest way to measure this is by its relative humidity (RH). The amount of water vapour the air could hold at that temperature (if water were freely available so that the air became saturated) is compared with what it is actually holding. This can be calculated by using a wet and dry bulb thermometer. Evaporation from the cotton gauze at the wet bulb causes heat to be taken from the adjacent thermometer. This causes a drop in temperature, similar to the effect of wearing wet clothes. The drier the air, the greater the rate of evaporation and the greater the difference in temperature between the wet and dry bulbs. Using a sliding scale, the difference in temperature allows the RH to be calculated.

Electronic humidity gauges are available for spot checks to see if conditions are right for the egg in both storage and incubation. Cheaper hair hygrometers, where the value is read straight from the dial, are quite effective. For a check on the accuracy of these hygrometers, use a wet and dry bulb hygrometer, which is unwieldy to use but accurate.

Humidity and Weight Loss in the Incubating Egg

The egg has to increase the volume of the air sac and lose about 13–15 per cent of its water by weight to enable it to hatch. Eggs are porous and so in an incubator at RH 40 per cent, eggs should lose water more rapidly than at RH 55 per cent. As the egg gradually loses water through its pores, the air sac at

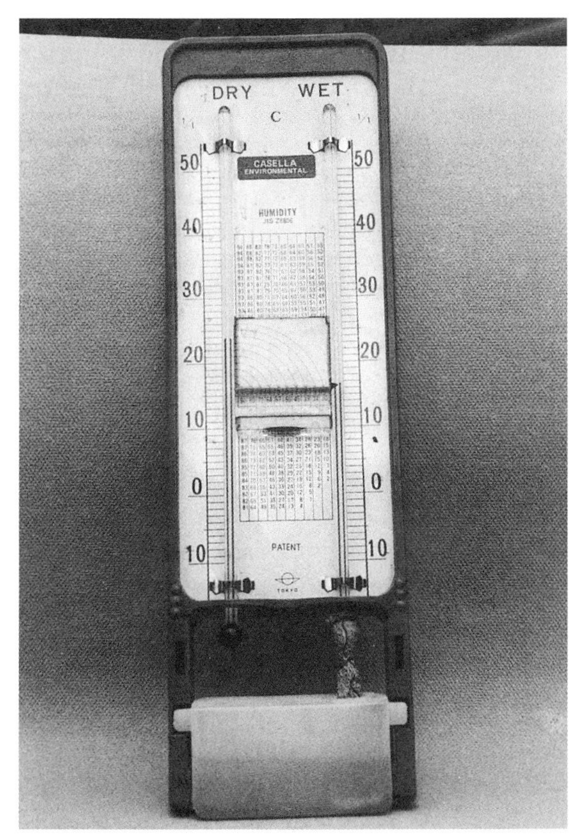

Fig 152 A hygrometer – wet and dry bulb thermometers. The dry bulb measures the air temperature. The wet bulb is sheathed in a damp cotton gauze, which wicks up moisture from a reservoir. Evaporation cools the wet bulb; the difference in temperature allows the relative humidity to be calculated. The reading from the instrument can be used to check the less accurate hair hygrometer. In the incubator, the bulbs should be at the level of the eggs for checking relative humidity at the egg temperature.

the bulbous end increases in size. This air space is needed to enable the duckling to break into it and breathe in air just prior to hatching and then to give it room to manoeuvre to get out of the egg. After that point, there is no air sac on candling.

Small incubators always arrive with a water tray and it seems obvious that waterfowl need high humidity for their eggs. The standard recommended humidity is 55 per cent. In a room's normal humidity of 65 per cent, water will generally have to be added to the incubator to achieve 55 per cent.

However, many early to mid-season duck eggs will not hatch if water is added to the incubator. As a general rule with still-air incubators, do not add any water unless the air sacs look enormous and excessive dehydration is a problem. This can happen with a cool, dry north wind in spring but the most common adverse condition in Britain is a spell of continuously wet weather, when RH in the house and incubator is too high. If there is water in the incubator, the ducklings will fail to hatch because there is insufficient weight loss, the air sac does not grow and they end up 'dead in shell'. With no water added, the average humidity often ends up just right.

With a room RH of 60–65 per cent, a dry incubator will run at about 38–45 per cent and this is often correct for early eggs. Too much weight loss is unlikely to happen until later in the season, in May or June, when eggs seem to dehydrate more rapidly because of their own changing properties. Depending on the breed, water may be needed in the incubator at this time of year to achieve 50 per cent RH in order to stop excessive dehydration. Note that these are only 'rule of thumb' guides and that candling and/or weighing the eggs is essential to monitor their progress. Forced-air incubators generally need more water than still-air incubators because moving air is more effective at evaporation than still air.

Variables Affecting Hatching

There are so many variables involved in the correct amount of water loss for hatching that it seems a miracle that some hatches achieve almost complete success. However, if some of the variables are understood, at least the incubator manager stands a better chance of success.

- Climate varies a great deal over the UK. An incubator operated in East Anglia where rainfall can be below 600mm requires a different input of water from one operated in the uplands of Wales.
- Air masses affecting the UK change frequently. Tropical maritime air from the south-west carries far more moisture (because of its warmth and passage over the sea) than polar continental air from the east, which carries little moisture in comparison.
- Microclimate affects humidity. Coastal areas affected by sea fogs in spring, and areas adjacent to bodies of water, experience higher humidity which may hinder weight loss.
- Egg shells vary in their porosity. This can vary between breeds, between individuals of the same breed and over a period of time in the same season. These are some examples:
  - if a mixed batch of eggs from various ducks breeds are put into the same incubator, they will not dehydrate at the same rate (*see* Fig. 162);

  - individual ducks of the same breed may have more porous eggs than their sisters' – this may be due to their rate of secretion of shell or variations that the birds are imposing on their food intake;
  - over the egg-laying season, eggs generally become more porous and lose water more easily – this can be remedied by adding water to the incubator, if required, later in the breeding season (ambient RH can be even higher in June than March).

Experienced incubator operators are generally able, without hygrometers, to estimate the conditions that work for their incubator in their region, for a particular spell of weather. They simply know from years of practice what certain conditions bring. Vernon Jackson, for example, using his old cabinet Hamer incubator, simply throws a bucket of water on the concrete floor of the incubator outhouse if he thinks the weather is too hot and dry. Vernon's maxim is that more eggs are ruined by too much water than by too little. There is no substitute for practical experience, and the best thing a new operator can do is to have a go and see what happens, following a few 'rule of thumb' tips.

- Do not add water to a still-air incubator during the main incubation period (day one to twenty-eight) during February to April. If the moisture content of the eggs is correct, additional water in the incubator is only needed after half of the eggs have pipped. Anderson Brown points out that it is helpful to ensure that the membranes are as thin as possible by encouraging evaporation, which also enables the ducklings to get at the air in the air sac. The hatchling does not get glued to the shell when it has only punctured a tiny hole; it is only when it sets about encircling the shell that high humidity is valuable to prevent dehydration of the membranes and sticking to the shell.
- Never add water in a wet spell of weather.
- Add water in a prolonged dry spell if the air sac of the eggs is looking large.
- Expect to add water during May–June if the air sacs are looking large. The eggs should be more porous.
- If eggs of certain breeds are hard to hatch, wash these eggs thoroughly in warm water and scrub with a nail brush. This removes their protective coating and makes them more porous.
- If there is an incubator and a separate hatcher, any eggs with a large air sac can be transferred (even several days earlier) to the humid hatcher to halt water loss whilst the wetter eggs can be left to dehydrate in the drier incubator. This is another advantage of having a separate hatcher and incubator.
- Monitor a sample of the eggs by weighing them to calculate percentage weight loss.
- If this seems too complicated, go for percentage hatching. Get an automatic cabinet incubator with a bigger capacity than a still-air incubator. Run it at about 40–55 per cent humidity (depending on the month and egg) and a fair number of eggs will hatch anyway.

The most valuable instrument to aid success in hatching is a good candler. With experience, it is possible to tell visually if the eggs are too wet or too dry with a good degree of accuracy. This is not as reliable, however, as monitoring weight loss and the eggs' response to the ambient humidity. This is partly because there can be a rapid change in the air sac at about day twenty-two to twenty-four when the embryo can 'drink' its own fluids and the sac can suddenly grow large and the embryo look very dry.

'Dead in Shell' – Insufficient Water Loss

Eggs that fail to lose enough water result in 'dead in shell' ducklings, i.e. they have tried to break the membrane to breathe, but have failed due to excess liquid. If, despite running the incubator dry, the air sacs are still not large enough, or the eggs are showing by weight loss calculations that they are too wet, then there are three possible courses of action.

- Put the eggs under broody hens in a dry nest made of chopped straw and shavings, preferably in a dry location in a shed. Broodies are very good at finishing off wet eggs perfectly as long as they have about a week.
- If the eggs have to stay in the incubator, open up the vent (if it is variable) at the base. This will improve the through-flow of air and dehydration. The eggs need more oxygen in the later stages of development too. Check that the temperature is correct when conditions are changed and run the eggs slightly on the hot side rather than cool to drop the RH further.
- During the mid-incubation period, try Pearce's recommendation (*see* page 145) of cooling the eggs daily.

Water Loss too Great – Spraying Eggs

A fine spray of warm, clean water once a day is often recommended for duck eggs close to hatching. The

Fig 153 In this Brinsea incubator being used as a hatcher, the water trays are full but the RH can only be raised to 75 per cent on the hygrometer by packing wet cloth in the grooves at the sides of the egg tray.

idea is that this prevents the internal membranes from being desiccated and impeding the duckling from escaping its shell. Note, however, than Anderson Brown recommended that thinning, dry membranes help the duckling pip, and I concur with his advice. Spraying with water is only good practice if the air sac is too large; if the air sac is ideal, leave well alone. Ducklings with correct moisture loss do not need excessive moisture for hatching, and filling the water trays when half the ducklings have pipped is ample.

To maintain a high RH for eggs that have dehydrated too much, spread a wet cloth over the unused portion of the incubator and then leave the incubator closed to keep the moisture in. It is the area of evaporating moisture that makes the biggest difference, and a warm, wet cloth at the level of the eggs (but not touching the eggs) on the floor of the incubator is the most effective method of raising the RH and useful during hatching. Note that evaporating water can initially drop the temperature of the incubator; it is essential to monitor it.

Candling and Development of the Egg

Candling involves shining a light through an egg to examine its internal structure. It is useful for several purposes.

- It saves incubator space by the rejection of infertiles early.
- It ensures that dead embryos, which cause rotten eggs, are rejected. It is essential that incubators are kept clean and that rotten eggs are not allowed to explode. If the incubator is in the house, bad eggs will be noticed before this stage anyway and can be removed before they become even more offensive.
- Finally, candling allows monitoring of the air sac and embryonic development and therefore the ability to monitor conditions for successful hatching.

Fig 154 A home-made candling device. This wooden box has an oval hole on the opposite side.

A torch can be used for candling but a good candling device is easier and quicker to use. Proprietary brands can be purchased, but if you have the time a simple one can be made much more cheaply out of a standard light fitting, plywood and a low wattage light bulb.

A newly laid egg shows a disc of cells, two layers thick, sitting on top of the yolk. This spot is 3–4mm across and, if it is fertile, has a raised translucent centre. The mesoderm forms in the middle within a

Structure of the Non-Incubated Egg

The egg is covered externally by a **porous shell** formed from the **external cuticle** and the chalky **testa**. This chalky layer gives the egg its strength and provides the calcium to be partially absorbed by the embryo.

Inside the shell is a **membrane**, which has two layers that separate at the broader end of the egg to form the **air cell** or **sac**. The outer membrane is attached to the testa and the inner membrane to the albumen. Gases are exchanged between the albumen and the blood supply of the embryo and the exterior through the porous shell structure

The **albumen** is held within the membrane and within this are two twisted cords, the **chalazae**. Their function is to suspend the yolk in the albumen. These cords are coiled in opposite directions so that when the yolk rotates in the albumen one coil is wound up as the other unwinds. Turning the egg in the same direction all the time will therefore destroy the egg's structure. Most of the albumen is formed from **dense albumen**, which extends to the ends of the egg where protein fibres are attached to the ends of the shell. The chalazae are within the thick albumen. The inner and outer part is **fluid albumen**, the inner fluid layer allowing the yolk to rotate on turning. The albumen is a store of protein and water and is used up during incubation.

The **yolk** is rich in fats and proteins and is surrounded by a thin membrane. The yolk itself consists of alternating light and dark layers, which were laid down by night and day, respectively. The centre of the yolk contains a ball of protein-rich yolk and a narrow column of this extends to the surface where it broadens into a cone. The **blastodisc** sits on top of this. The majority of the yolk is not used during incubation; it is drawn into the abdominal cavity of the duckling just before hatching

Whatever position the egg occupies, the **blastodisc (germinal disc)** occupies the upper position. This is a small, white spot (about 3mm across) from which the duckling will grow. The part of the yolk next to the germinal disc is less dense than the rest of the yolk so it tends to float upwards, rotating the yolk so that the germinal disc is always on top. In the early stages of development the embryo can only use the nutrients in contact with it; turning the egg gives it a new source of food and oxygen from the inner liquid albumen.

The yolk and the germinal disc originate in the ovary. The albumen, shell membrane and shell are formed by the oviduct.

(Adapted from Charnock Bradley, 1950; and Brooke and Birkhead, 1991)

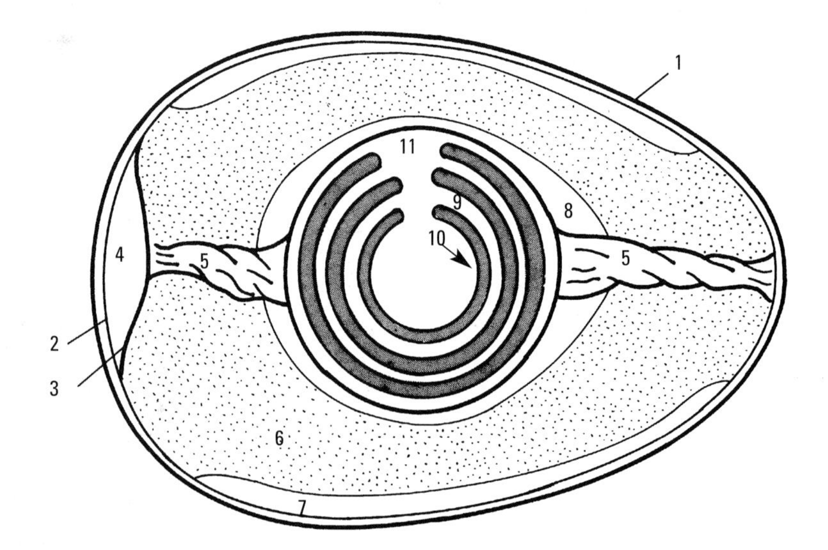

Fig 155 Structure of the non-incubated egg: 1 shell; 2 shell membrane; 3 egg membrane; 4 air sac/cell; 5 chalazae; 6 dense albumen; 7 outer liquid albumen; 8 inner liquid albumen; 9 white yolk; 10 yellow yolk; 11 blastodisc/germinal disc.

few hours of laying. On incubation, the neural tube is formed and then the brain and spinal cord. Initially, the embryo obtains sufficient nutrients through its lower surface but as it grows it creates four membranes: the yolk sac, amnion, chorion and allantois.

After three days of incubation, the disc is 25mm across and a network of blood vessels is forming to carry nutrients from the yolk. The blood is already pumped by a primitive heart. An experienced operator can recognize fertiles at this stage. By the sixth day, most of the organs are taking shape and there is an extensive network of veins. The eye is particularly well developed and can be seen as an obvious spot on candling. There is no mistaking a fertile duck egg by these features.

Eggs that are infertile will not show a 'spider' of veins. Eggs that have started then died will not show an eye. Also, the degeneration of the veins after the death of the embryo shows a 'blood ring' instead of the 'spider'. Sometimes there are black patches inside a transparent egg but no veins. These are probably bacteria colonies and the egg needs removing.

Candling should not take place unnecessarily as the heat from a light bulb is excessive and a rapid inspection is best. However, the eggs will need checking again. As incubation progresses, the whole of the egg will darken between day ten and fourteen. There should be a clear dividing line between the membrane of the air sac and embryo. An indistinct line and a pale band adjacent to the line indicate that the egg has died. The egg is also cooler to the touch than the other eggs. Eggs that have started and then died are the dangerous ones that will go rotten and could explode, infecting the whole incubator and its contents.

Development of the Air Sac

The size of the air sac only changes gradually up to eighteen days, even though there has been a fairly steady loss in weight, but after that regular and more rapid change should occur visually. On candling, mark the extent of the air sac with a pencil line at, and after, day eighteen, so you can see how it is developing. Usually the eggs with the best development of this feature are the early hatchers by as much as 24–36 hours.

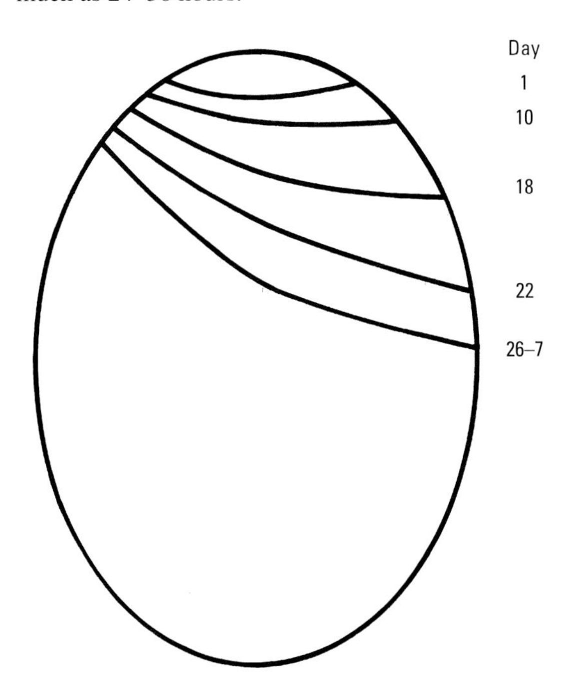

Fig 156 Ideal air sac development.

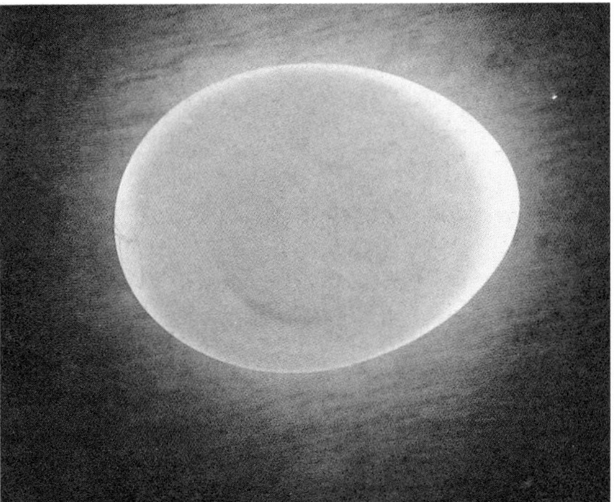

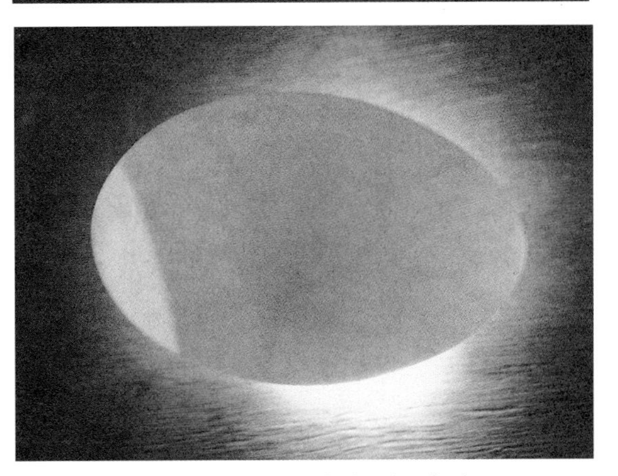

Figs 157–159 Candling eggs during incubation: development of the embryo. (top) An egg after four days of incubation showing the typical 'spider' network of veins. (centre) A fertile egg that has started to develop and then died, also photographed at four days. It shows the typical 'blood ring' of veins which have deteriorated. (bottom) A healthy egg at eight days. The darker central region is the head and eye. The air sac has already grown.

Sample of eggs from April 2000 to show how more water was needed in the forced-air incubator

Forced-air incubator: 3 April to1 May 2000:
- all water trays filled day 1, day 5, day 10; RH varied from 40 to 50 per cent
- water trays kept full days 11–25; RH 50 per cent
- eggs transferred to still-air incubator day 25–28; RH 55–75 per cent when sprinkled with water; water trays full.

Day	1	5	10	15	20	25	Actual Wt Loss (g)	Wt Loss (%)	Target Wt
White Runner	91	88	84	80	78	76	15	16.5	77.5
Rouen	87	85	83	81	80	78 *	10	11.5	74
White Runner	76	72	69	66	64	62	14	18.4	64.5
Fawn-and-White Runner	71	69	66	63	61	60	11	15.5	60.5
Orpington	70	67	64	62	60	59	11	15.7	59.5

*Retained in forced air day twenty-five to twenty-eight to dry.

Eggs at 15–16 per cent loss hatched the best: 100 per cent of Trout, Fawn-and-White Runner and Orpington. White Runners were rather dry but seven of nine hatched; only one of three Rouen hatched.

Sample of eggs from May 2000 in a still-air incubator

Still-air incubator: 27 April to 22 May 2000:
- No water added, incubator typically running at RH 45 per cent days 1–25
- Water added day 25; RH 75 per cent.

Day	1	5	10	15	20	25	Actual Wt Loss (g)	Wt Loss (%)	Target Wt
Rouen 20	90	88	86*	84	82	80	10	11.1	76.5
White Runner 18	84	80	77	73	69	67	17	20.2**	71.5
Fawn-and-White 15	72	70	68	65	61	60	12	16.7	61.0
Trout Runner 15	69	67	64	61	58	57	12	17.4	58.5

*Removed to drier incubator – Rouen eggs difficult to dehydrate this year.

**White Runner eggs dehydrated too much, but still hatched.

If the air sac is too large (perhaps because the weather has been dry or because of late season eggs) then the humidity needs raising. If the air sac is too small, then RH has been too high. Humidity is difficult to regulate down in still-air incubators, particularly if you have not had water in the water trays anyway.

Monitoring Weight Loss

Although the size of the air sac is a useful visual guide to check for correct dehydration, it is not as useful as weight loss. Eggs must lose about 7 per cent of their weight half-way through incubation and 14–15 per cent by day twenty-five to twenty-seven. Duck eggs are small compared with goose, so an accurate balance is needed for individual weighing.

- Weigh a sample of marked eggs, of different breeds.
- On day fourteen (or at five-day intervals, if you have the time) weigh the same eggs again. They should have lost 7 per cent of their original weight, i.e. (weight loss/original weight) × 100 = 7 per cent loss by day fourteen.
- If the weight loss is 7 per cent, continue incubation under the same conditions as days one to fourteen. Compensate the humidity if the eggs are either too wet or too dry. It is preferable to have eggs on the dry side rather than wet.
- If a particular breed's eggs are too wet (or too dry) these will have to incubated under different conditions if the hatch rate is to be good. The target weight loss can be plotted on a graph and the eggs' progress monitored against this (*see* the graph opposite).

Hatching in the Incubator

Cleaning

It is preferable to keep one incubator as a hatcher and to give it a thorough clean every time it is used.

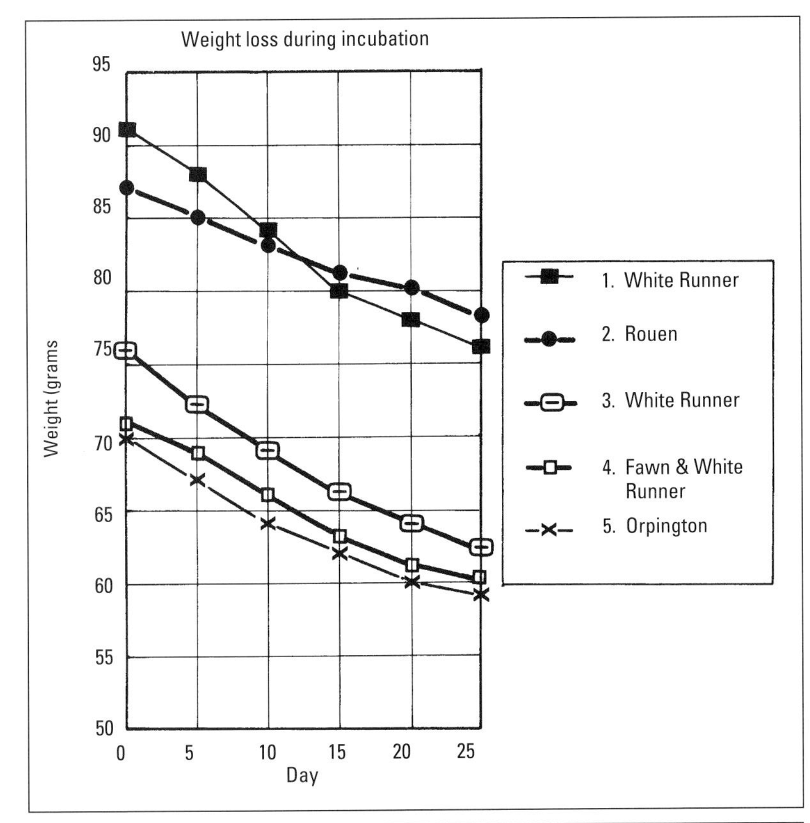

Fig 161 Sample of eggs from April 2000 in a forced air incubator (see table opposite)

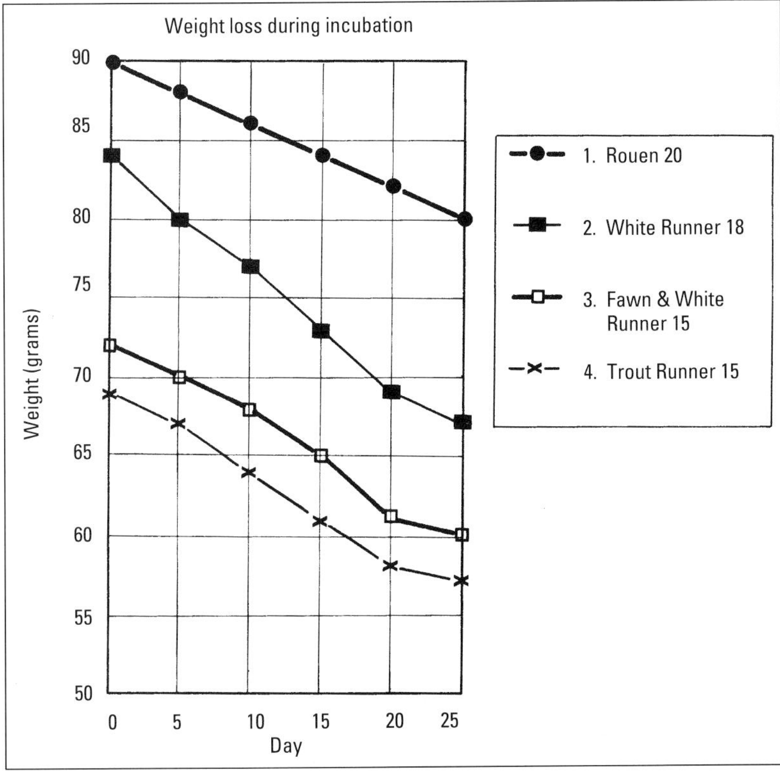

Fig 162 Sample of eggs from May 2000 in a still-air incubator (see table opposite)

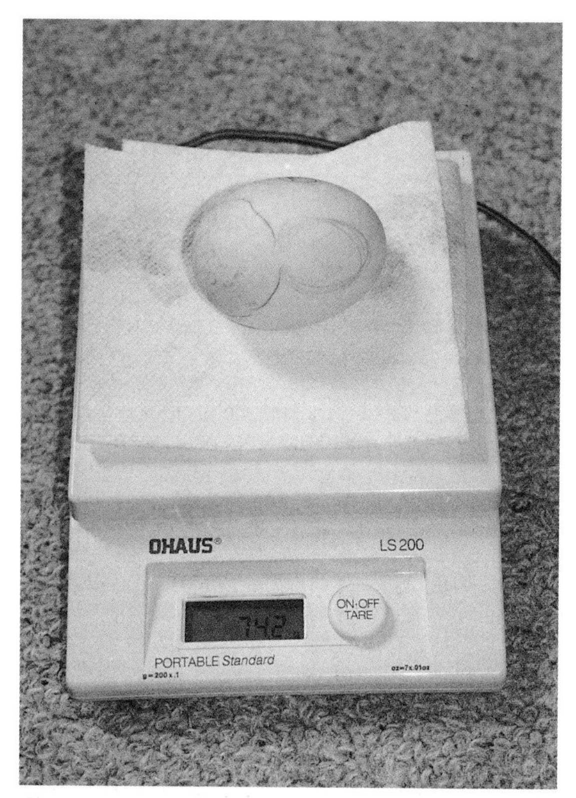

Fig 160 Weighing eggs: an accurate balance is needed.

It can, of course, be used as an ordinary incubator in-between hatches. The hatching incubator can be a smaller model than the main incubator. The pre-requisite is that is must be easy to clean. It is best if the whole of the base of the incubator can be immersed in the sink and scrubbed with detergent. Milton can also be used afterwards as a disinfectant, or a sanitizer such as Virkon. The interior of the lid should also be wiped down with a damp cloth and disinfectant.

Suitable Hatching Materials

Rolls of hatching paper are available from incubator stockists and are useful because they help to keep the incubator clean. Perforated aluminium and wire mesh are harsh on the raw umbilicus of a newly hatched duckling, so special paper or textured paper kitchen towels can be used. Newspaper is no good because it is too smooth and will only encourage spraddled legs.

Regulating the Temperature

The temperature of the hatcher will need watching carefully. Active hatchlings generate a lot of heat and the incubator temperature will therefore rise to unacceptable levels if you do not regulate the control downwards. Do not allow the hatcher to exceed the core egg temperature of 37.3°C. Eggs that are active will feel very hot and may be 2°C higher than their recommended core temperature because of their exertions. They feel far too hot and should be allowed to cool. Those having a rest (or dead) will feel cooler. On the whole, it is better to have the air temperature of the hatcher on the cool side at 36°C rather than on the lethal hot side. Go by the feel of the eggs as much as the thermometer.

Pipping and Oxygen Demand

Throughout incubation, the developing egg has consumed oxygen and released carbon dioxide. The vents in the incubator have been designed to exchange the gases by the correct amount, as long as the incubator room is ventilated. Oxygen demand increases as the egg grows, so if there are variable vents on the incubator it is a good idea to open these near the end of incubation when the eggs are also warmer, because they are partly maintaining their own temperature.

Close to hatching, the embryo has to change from allantoic respiration to using its lungs. When the allantoic circulation begins to shut down, the rise in carbon dioxide causes the embryo's neck muscles to twitch. The beak jabs upwards and pierces the allantois and then the air sac, allowing the duckling to take its first breath into its lungs. The beak tip has an 'egg tooth', which is quite noticeable on hatching, but drops off shortly after.

As the neck muscles twitch, the yolk sac is also withdrawn into the body cavity. The twitching also breaks or 'pips' the egg shell so that the bird has access to fresh air and, once this is achieved, it often rests. This pause varies in duration but during this time the yolk sac should be fully withdrawn and the blood vessels of the egg are shut down. This whole process of switching respiration takes about two days (Anderson Brown, 1979).

Hatching

Once half the ducklings have pipped (or earlier if the eggs were too dry), fill the incubator's water trays. Allow the humidity to build up (leave the incubator alone) but position the thermometer so that you can monitor the temperature easily.

After the customary rest for 24–36 hours after pipping, the ducklings should rotate in their shells and complete the hatching process. They tend to be slower at this than chickens, but faster than goslings. It is worth while keeping an eye on them as a few unfortunate individuals do this then fail to flip off

Fig 163 The duckling pips the shell at the bulbous end, then rests for abut 24–36 hours (right). It then rotates, pushing with its muscles, until it a has completed a circle (left). Occasionally, the duckling does not flip the lid open, and it should be helped out at this stage.

the lid and suffocate. Do not interfere with a duckling unless it has almost completely rotated around in its shell. It will not have resorbed all of its blood supply, and it will lose vital fluids. Also, if left alone, moisture is retained within the shell and the duckling is less likely to get stuck by dehydration and gluing to the shell.

It is sound advice to leave the incubator alone while hatching is taking place. Humidity actually builds up because the ducklings are wet and slimy when they hatch. However, at some point the incubator seems full of heaving ducklings that have hatched early and are trampling everything else, and it is a good idea to remove the early hatchers, put down a clean layer of paper, and check the unhatched eggs.

Late Hatchers and Non-Hatchers: 'Dead in Shell'

Once the majority of eggs have hatched, there may be eggs left that have not pipped at all, and others that have pipped but got not further. Sometimes these eggs are delayed because they were older eggs when they were set, or were the very last laid and rather wet. It is worthwhile giving these a little longer (29–30 days). However, it is inadvisable to give too much help to late hatchers; they frequently have something wrong with them and do not do well if they are helped out.

Eggs that have developed to quite a late stage and then failed to pip are known as 'dead in shell'. Development may have been retarded by too low a temperature, or by the egg being too wet because of lack of dehydration. If they die at this stage, it is worthwhile checking the dead hatchling to see if there is excessive moisture, which will run out if the inner membrane is torn open. Usually in such a case the yellow yolk sac remains attached to the umbilicus externally, so the bird would have stood a poor chance of survival. If there are several eggs like this, the incubator has been run at too high a humidity. Always check to see if eggs are dead by candling or breaking into the air sac first; this way any late duckling is not damaged.

Deformities

Eggs that take a long time to hatch frequently have something wrong with them. Retarded development is one cause but so is deformity in the duckling. Twisted legs or a crooked neck make it more difficult to hatch. These are likely to be genetic deformities, but occasionally they disappear in a few hours and seem to have been caused by malpositioning. If there are several ducklings with such a deformity, it is either a genetically acquired characteristic or possibly vitamin deficiency, which is well documented in chickens but not ducks.

Malpositioned Ducklings

A few ducklings pip at the wrong end of the shell. They should break into the air sac at the bulbous end, but some attempt the pointed end. These ducklings may need help because there is no air sac for them to breathe easily, so they may suffocate. Cautiously clear a small air space for them to breathe, even if they bleed; they tend to bleed more at this end of the shell. They do not have enough room to manoeuvre and perform the can-opener action of correctly positioned hatchlings. Consequently they may need help to hatch. Leave them alone except for checking the air supply and only hatch them when other birds, which pipped at the same time, have hatched themselves normally, and the membrane is no longer bloody. The strongest ducklings will still hatch themselves.

This condition is alleged to occur if the eggs have been set with the air sac end too low, but it seems to still happen when the eggs have all been set in the same normal way. To guard against malpositioning, always handle eggs slowly and gently, and avoid vibration and jarring. Turn eggs as frequently as possible.

Checklist

Incubation is as much an art as a science, and there is no substitute for observation and experience. If you are successful first time, it may not last, as there are so many variables over which one has little control. If you are unsuccessful, run through the following points.

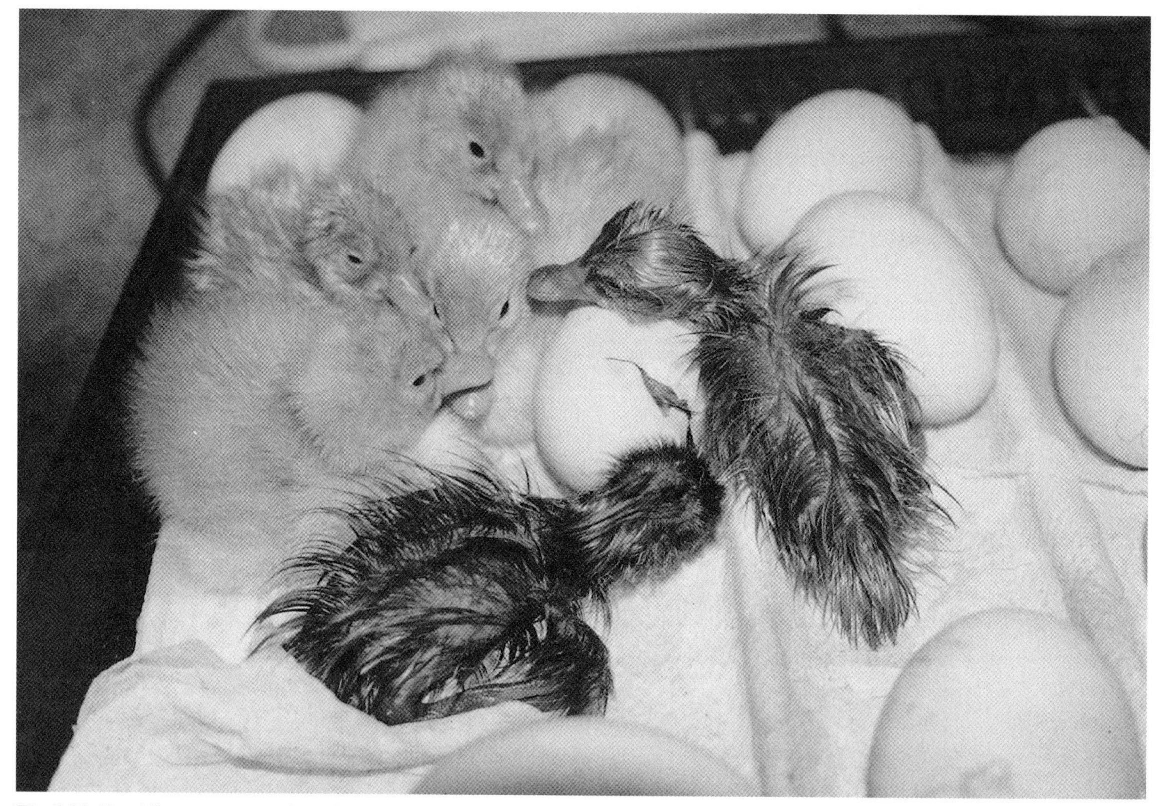

Fig 164 Ducklings are wet when they hatch and it is important to make sure they do not get chilled.

Infertile Eggs

- Old eggs set; use eggs ten days old or less.
- Ruptured membrane – pre-warming neglected.
- Drake infertile: behavioural problems; fighting with other drakes; bad feet; overweight; dropped/ withered penis.
- Poor diet of breeders, e.g. insufficient vitamins – duck and drake.
- Congenital defects, especially from very in-bred birds carrying lethal genes.
- Poor storage and handling of eggs.

Bad Eggs

- Eggs dirty when collected, especially from wet, muddy areas.
- Eggs washed in cold water, germs drawn in.
- Cracked eggs set, bacterial infection.
- Dirty incubator, bacterial infection.
- Embryo has started to develop and has died.

Embryonic Mortality Before Day 26

- Incorrect storage, e.g. too warm, some embryonic development prior to incubation.
- Insufficient turning.
- Fluctuating temperatures in incubator, embryo dies.
- Poor diet of breeders, embryo dies.
- Congenital defects.
- Infections (*see* above).

Dead in Shell: Mortality in the Last Two to Three Days of Incubation

- Early season eggs; things may improve later in the season when eggs are more porous.
- Humidity too high; insufficient air sac – liquid runs from torn membrane, yolk sac not absorbed.
- Humidity too low – air sac very large.
- Malpositioning of duckling – chance factor or incorrect storage of egg.
- Deformed duckling – congenital defect.
- Overheating – a lot of heat is generated by hatching eggs; watch the air temperature of the incubator.

Early Hatching

- Incubator temperature too high; ducklings may have bloody umbilicus – insufficient time to resorb blood from veins of egg. Occasionally, ducklings hatch themselves and bleed profusely.
- Smaller eggs of young ducks or smaller breed (Calls and Appleyard Miniatures) – no problems.

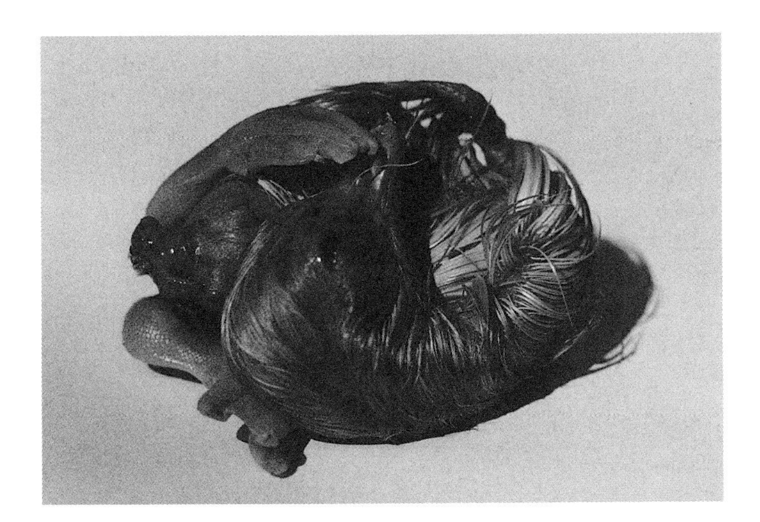

Fig 165 This Call duckling failed to hatch; it was 'dead in shell'. The yolk sac had not been resorbed and the duckling may have died at day 24. Calls are more difficult to hatch than the other breeds and more often dead in shell. The egg was slightly wet but water loss can be correct and the Calls still fail to hatch. Call hatchability is extremely poor in still-air, hand-turn incubators. An incubator with a well-regulated temperature and automatic turn is needed, but will still not beat natural hatching.

Late Hatching

- Old eggs, or last egg set (which has lost less water).
- Large eggs can take up to thirty days, but twenty-eight to twenty-nine is ideal.
- Incubator temperature too low.
- Deformed duckling.

Deformities

- Inadequate diet of breeders – mineral and vitamin deficiency.
- Crooked toes and turned ankles can result from too high temperatures, vitamin deficiency or inbreeding.
- If eggs from the same bird also produce deformed ducklings when hatched under broodies, this is a genetic defect and the same pair of birds should not be used for breeding.

8 Rearing Ducklings

If a few simple guide-lines are followed, ducklings are easier to rear than chicks or goslings. They must, however, have the correct food, a suitable water container, be kept dry and clean, and be brooded at the correct temperature. They must also be protected from vermin. If there are only a small number of ducklings, it is possible to improvise and adapt equipment that is to hand but, for more than ten ducklings, it is worth buying purpose-made equipment to cut down on work and worry. Get a catalogue from a good poultry supplier (*see* Useful Addresses) to see what they stock.

The best way to start off a small number is to keep an eye on them in your house for a couple of days. During this time they will have become very familiar with people, which is important in keeping them tame. Hand-feed them and get them used to being picked up.

Whether the birds are for pets or commercial production, accustom them to your voice and movement. Always talk to the ducklings when opening doors to let them know that you are there. Spend some time on feeding them and get them used to your hands bringing food and water. This prevents the ducklings from rushing into a corner and trampling each other. This is as important in commercial production as in a hobby.

Transferring from the Hatcher

Leave ducklings in the hatcher until they have dried out and fluffed up. As they dry, the protein coating that sheathes the down will fall away, leaving yellow 'incubator fluff' behind. If they are not crowded, ducklings can be left in the hatcher for 24–36 hours, as long as the temperature is turned down from 37 to 35–36°C when they are dry. Check that they are comfortable. The ducklings should then be transferred to a ready-warmed, draught-proof box, without chilling them.

Early Problems

Occasionally, almost the whole of a hatch of ducklings can die, often at about one week old. This may happen to only one breed in a batch of several breeds that are otherwise healthy. The problem is probably caused by inbreeding, though sensitivity to a particular cocciodiostat in chick crumbs is a possibility.

When ducklings have hatched well, they are generally free of problems. Late hatchers are more likely to have something wrong with them, and also to become ill. They may be all right for a few days, but then lose their appetite. On close inspection they often have laboured breathing or a whitish discharge from the vent, which is sticky and quite unlike the other ducklings that are clean. The laboured breathing is often caused by lung congestion which Cook (1899) remarked upon, and the white discharge by a bacterial infection from the yolk sac. Such ducklings should be isolated from the rest as soon as possible and are best put down. They can occasionally be saved by antibiotic treatment if caught early, but never grow up into strong birds. Keep a close eye on ducklings for these problems during the first couple of weeks. Once they are out in clean grass runs, these conditions should not occur.

Heat

Heat Lamps

If the ducklings are reared in small batches as pets, they can be kept warm enough by a kitchen stove or in the airing cupboard overnight (as long as the temperature is monitored) but it is often better to use a heat source from light bulbs or a heat lamp. Two 40-watt bulbs suspended above the duckling's box will emit sufficient heat in a well-insulated box kept in a warm room in the house. After a couple of days, the amount of food and water they are consuming will be rising dramatically and they will need appropriate equipment and bedding in an outhouse.

In an outhouse, a bigger heat output is required. Glass infra-red bulbs are cheaper than ceramic bulbs and feel as if they give out more heat for the same wattage. The fitting for the bulb must be ceramic to stand up to the high temperatures. Unlike goslings and chicks, ducklings do not seem to be so badly affected by constant ruby light, as long as the light is raised as they grow. White light from the clear lamps is not recommended; it destroys vitamins in food even faster than ruby lamps.

Ceramic bulbs, which do not emit light, allow the ducklings to settle to sleep overnight. The heat element lasts a long time and so the ducklings are unlikely to be lost by chilling because there is less frequent breakdown. These ceramic lamps are also less likely to shatter than glass ones if splashed with water. They are more expensive than ruby lights but more reliable.

The ducklings' drinker and food must be positioned away from any heat source because the heat, and light from glass bulbs, also destroys vitamins. This will result in weak legs and rubbery beak in extreme conditions. The cool area of the box also gives the ducklings a place to escape to if the height of the lamp is incorrect and far too hot.

It is preferable to have two light sources (especially in an outhouse where temperature fluctuations can be considerable overnight), in case one lamp breaks down. If ducklings are kept in different groups with a partition between them, a back-up lamp can be shared between two batches, whilst their main lamp heats the majority of the area.

A new type of electric brooder lamp, which looks more robust and reliable, is now being advertised by Maywick, as well as their gas brooders. This type may well be worthwhile investigating.

The height of the lamp determines the temperature on the floor. The floor temperature should be measured with a thermometer, but the behaviour of the ducklings is also a good guide. If they are in a huddle under the lamp, they are cold. If they are spread out in two or three sleeping groups, they are happy. Ducklings are less prone to crushing each other when sleeping in huddles than goslings, but commonsense must be exercised in deciding how many ducklings can be accommodated by one lamp. As a general rule for the UK climate, two lamps give up to fifty young ducklings enough space. A batch of twenty-five has enough floor space with one lamp but still needs a back-up lamp.

As well as monitoring the floor temperature under the heat lamp or brooder, it is important to make sure that the area is free from draughts. This might be difficult in an outhouse and it may be necessary to give the ducklings a cardboard box (turned on its side with the flaps cut off) to find a warm, draught-free area. If a fairly large space is being used for the ducklings to grow into, they must be confined to a smaller area to start with, otherwise they can stray away and become chilled. A circular surround about 17½in (45cm) high is useful for larger numbers because it stops the birds crowding in corners and suffocating each other. As a general rule, for rearing on a larger scale, the temperature of the floor of the house at 6½ft (2m) from the lamps should be about 9–12°C cooler. Excess heat is bad for ducklings and they must have a space where they can choose the correct temperature.

Fig 166 Two groups of ducklings of different ages share the back-up heat lamp. The partition between them must be made of non-flammable material or be well away from the heat source. The total floor space of the two compartments is 1m² of wire mesh and it can accommodate up to fifty ducklings at three days old, though more space is preferable. They will stay here until two weeks old when the floor space is opened up to 2m² for up to fifty ducklings. Fairly crowded conditions to start with do not seem to adversely affect the birds. They do as they get heavier.

As the ducklings grow, they get nearer the heat source but they also require less heat, so the lamp must be raised by 1–1½in (2–3cm) every two to three days. Monitor temperature and behaviour. The floor temperature should be gradually reduced to 'harden

Recommended temperatures for growing ducklings.
Ducklings outdoors from 2–3 weeks do not need these temperatures all day, but they should be maintained overnight. Such temperatures may be achieved by the body heat of a group of ducklings in a coop.

	Heat required under lamp or brooder		Floor temperature 2m from lamp	
	°F	°C	°F	°C
Floor temp. first 7 days	90–95	32–35	70	21.1
Week 2	85	29.4	65	18.3
Week 3	80	26.6	60	15.5
Week 4	70–75	20.1–23.8	60	15.5
Week 5	60–65	15.5–18.3	55	12.7

*Fig 167 A traditional method of rearing ducklings and chicks on a shed floor was to use a brooder heated by paraffin. These are very effective, but also carry a greater risk of fire than well-maintained electric appliances. (*The Feathered World*, 1928)*

them off' for transferring to an outdoor coop and run. The age at which this happens depends upon the time of year, the weather and the number of ducklings to keep each other warm in the coop. The ducklings will need heat, however, for at least two weeks. In a well-designed coop with adequate bedding, ducklings can be transferred outdoors at two weeks old. This is as long as the weather is warm and dry. They may need heat up to four weeks in early spring; a temperature of 15–18°C is still needed to keep them comfortable but a huddle of ducklings in a coop will maintain this temperature.

Water

Food and water is not needed for the first 24–48 hours but, when reared under a heat lamp, water is often appreciated. If the ducklings are in a container in the house for a couple of days, limit the amount of water they have by providing a ceramic egg cup, of the cylindrical type (or a similar container). This will be stable because it has a broad base. Do not use a conventional fountain drinker initially because the ducks will become adept at siphoning out the water. They will also jump in the water and get too wet, and will end up cold and miserable.

When the birds are a few days old and eating more, they will need larger quantities of water and the drinking fountain is ideal. However, it must be placed on a grid (*see* below) that allows water to drain away, otherwise the ducklings' bedding will become a cold, soggy mess. A poultry drinking fountain is ideal to start with because the rim containing the water is shallow and narrow, discouraging drowning. It is deep enough for the ducklings to wash their beak and nostrils at this stage.

Fig 168 These Appleyard Miniatures, Calls and Trout Runners keep each other warm between two and four weeks in these old 'hay-box' style coops. The ventilation has been wired over to exclude predators such as Little Owls and mink.

Fig 170 Newly hatched ducklings are only given a small water container. This limits their water supply at first to keep them dry and warm. The plastic box for a small number of ducklings in the house is useful to start with, as it prevents leakage from their water supply.

Fig 169 (left) A drinking fountain for ducklings aged two days to two weeks old. (right) A larger drinker with a deeper rim for two to four week olds. The rim is too large for small ducklings, which could get lodged in it and get cold and die, or drown. The legs can be folded away if necessary.

Fig 171 An automatic system in a commercial unit: the flow of water into the drinkers is regulated by a polystyrene float so that the water, supplied by hose pipe, never runs out. Spillage runs through the weld mesh so that the ducklings stay dry and clean.

By two to four weeks of age, a larger quantity of water to wash the bill and eyes may be needed. There are larger-rimmed plastic drinkers designed for ducks, but a solid pie dish filled with water can be offered in addition to the poultry drinker. By this age, if they get in the water, the ducklings are old enough, and knowledgeable enough about the heat source, to avoid chilling. They are also beginning to use their own preen gland to oil up their fluff, just like adults. The pie dish must be shallow enough for the ducks not to be trapped inside it, chill and drown. It also needs to be on a grid for the water to drain away. Always be careful when replacing drinking fountains in the rearing area. The ducklings are often enthusiastic about getting clean water and can rush under the fountain and get crushed as it is put down.

The water must never be allowed to run out if dry food is available. The ducklings will continue to eat, even in the absence of water. When they are given water later on, the dry food swells up inside them and they can suffocate from the internal pressure. If food and water have run out, always give them water first. This should not be ice-cold; if they are desperate for drinking water, serve it lukewarm.

Give the water container a thorough scrub every day. It must be large enough to keep the ducklings in water for the period of time you are not around to fill it. Best of all, use an automatic system.

Bedding and Containers

When they have just hatched, ducklings do not make any demand on food and water, and are best kept in the house in a large cardboard box below the heat lamp. This can be lined with an old T-shirt or textured hatching paper. The layer can be topped up with clean paper or chopped straw as it gets dirty and, after one or two days, the ducklings can be transferred to a shed and the old bedding burned. Keep the bedding relatively dry by using only a small water container that is topped up fairly frequently. Ordinary newspaper can be used for bedding when the ducks are well on their feet. It is not suitable to start with as it will cause the ducklings to slip and get spraddled legs. A plastic container, such as one used to contain plants or the base of a hamster cage, is also useful instead of a cardboard box.

When the ducklings are transferred to a rearing shed, the first requirement is that it must be rat-proof. Rats seem to prefer ducklings and goslings to other types of food. Even if you think there are no rats about, the area will not stay rat-free if food is left available for them. The shed itself must be constructed of sound wood if it is to keep vermin out. Rats gnaw through rotten wood very quickly. Brick sheds with a concrete floor, checked for any gaps, are far safer.

Wood shavings are no good to start with because quite a lot will get eaten. They can be used when the ducklings are a bit older and trained on duck food. Hemp (used as an absorbent bedding for horses) should not be used. It swells up when it is wet, and kills the ducklings if they eat it. Hay must not be used either because aspergillae (moulds) develop rapidly.

Chopped barley straw, which must be free from mould, is best. Finely chopped dust-extracted plastic-baled straw such as *Sundown* is good because it is very absorbent. It may seem expensive but a little goes a long way and for small quantities of ducklings it is ideal. As it gets wet and dirty, the straw can simply be topped up into a deep bed. Keep the drinking fountain on a piece of raised weldmesh to stop straw clogging up the rim. If there is a slant on the floor of the shed, make sure that the water drains away from the bedded area, or place the fountain on a container to catch the spillage. Eventually, moulds and ammonia build up in the bedding and become a health hazard to the ducklings, which are in very close contact with the floor. For this reason, the healthiest way to rear ducklings, especially in larger numbers, is on wire floors.

Rearing on Wire Floors

Duck-rearing units, which consist of a box on legs, containing a wired area as a floor, can be purchased but they are not very large. If you have greater numbers of ducklings, it is best to construct your own rearing area from weldmesh and wood. This area (Fig. 172) has been constructed from panels of weldmesh stapled to planed wood. The panels rest on supports about 15½in (40cm) from the floor and the area of mesh is divided by thin plywood partitions which can be scrubbed. The wooden strips are not ideal as they collect droppings, and a system where the weldmesh is stapled to upright partitions

Table to show minimum floor space required for rearing ducklings

System	Maximum Stocking Rates
Metal mesh floors:	
Day-old to ten days	Fifty ducklings/m^2
Ten days to three weeks	Twenty-five ducklings/m^2
Three to eight weeks	Eight ducklings/m^2
Not advised for rearing pure breeds and breeder birds after three weeks; commercial ducks only	
On solid floor with litter:	
Day-old to ten days	Thirty-six ducklings/m^2
Ten days to three weeks	Fourteen ducklings/m^2
Three to eight weeks	Seven ducklings/m^2
In grass runs:	
Three to eight weeks	2,500 ducklings/hectare (twenty-five ducklings/100Sq m)*

*In well-grassed runs this density can be doubled. Note that the grass should be clean and the area rested from adult ducks before it is used for rearing. These figures are for commercial ducklings; pure breeds should not be reared in such numbers, and need more space to grow well.
(Figures taken from MAFF publication *Ducks*)

Fig 172 Weldmesh panels allow water and droppings to fall through, keeping the ducklings dry and clean. Ducklings sit on the paper mats when they are not busy eating and drinking. These heat lamps are ceramic, and more reliable than ruby glass, so a back-up lamp is not so crucial. The metal hoods of the lamps should not touch the metal sides of the shed.

Fig 173 Purpose-built rearing unit at Richard Waller's for large numbers of ducklings: gas brooders are preferred to electric. This type of brooder necessitates a good ventilation system. The ducklings have a continuous supply of water and also food from the tube feeders. The rearing unit is raised so that droppings and spillage pass through the mesh and can be cleared out from beneath (automatic drinkers and tube feeders are supplied by Bird Stevens and Co. Ltd).

is easier to keep clean, but not as strong.

The weldmesh itself gives the ducklings a good grip for their feet, preventing spraddled legs. It allows droppings and water to fall through, leaving the ducklings clean and dry. This system works well if the area is large enough to position the heat lamp away from the water and food containers.

The ducklings should also be given a mat under the heat lamp and just away from the heat lamp so that they can sleep comfortably. Mats can be any material that is readily cleaned, such as rubber car mats or carpet tiles that can be hosed and dried, but several sheets of newspaper provide a warm, absorbent surface, and can be renewed. This system works well for ducklings but not for goslings, which are too heavy on the mesh surface. They tend to gets their hocks caught in wider mesh (necessary for the larger droppings) and also try to eat carpet tiles and newspaper.

The waste material is wet and forms a slurry, which is best absorbed by sawdust or any other dry litter; it can then be regularly shovelled out. It should not be left to accumulate for long because blue moulds develop on the surface, and may cause lung infections in the ducklings. Hose down the concrete surface periodically, if possible. If the shed design does not allow for this, throw fresh sawdust down after clearing out the muck, as this seems to inhibit mould by drying off the area. Sawdust can also be used as a dry cleaning material. Scrub it into the floor with a stiff broom and then remove it. Start with clean sawdust again for the next cycle.

Fig 174 At two to four weeks, depending on the weather, the commercial ducklings at Richard Waller's are moved on to a veranda combined with an indoor rearing area. The water troughs at the outer edge of the veranda have the inflow regulated by a ball cock; the water is therefore ad lib, and the ducklings can immerse their heads. Waste water and droppings fall through the weld mesh of the veranda so the ducklings stay clean. If they are cold, they go inside via the pop-hole to a warm area of deep litter, which stays dry. The ducklings are therefore acclimatized before they go outdoors in the day time, after four weeks.

This system can be used until the ducklings are three weeks old, if it is too cold to put them out. In bad conditions, they can be kept in for even four weeks but it is particularly inadvisable to keep large ducks such as Rouen and Indian Runners indoors any longer. If the birds cannot go out, the heat lamps must not be on them directly, and vitamin supplements should be used. The birds definitely benefit from daylight, fresh air, green food and exercise. They will develop signs of vitamin deficiency, such as rubbery beak, enlarged hocks and bowed legs, if kept under poor conditions, and develop vices such as feather picking. It is important to get the birds out onto grass runs.

Before the next batch of ducklings are transferred from the hatcher to the rearing shed, the wire mesh rearing system must be thoroughly cleaned. If there have been no problems with infections in rearing the ducklings, scrubbing and hosing down the weldmesh and partitions, and air drying them in sunlight, is adequate disinfection for small-scale home production. If there have been problems, try Barrier liquid, which is effective against a range of bacteria affecting poultry, yet is a natural non-toxic product. Virkon is also an effective bactericide.

Always operate a one-way system, i.e. move the youngest birds onto clean premises; never put a batch of older birds back into premises that will be occupied by young ones afterwards. The only exception made to this rule is where a duckling may be kept back rather than moving it out to a coop because it is too small (but otherwise seems healthy).

Food During the First Five Weeks

The First Two Days

Ducklings are quickly on their feet and interested in food, but they are best not fed for twenty-four hours and can last forty-eight hours with no food. The yolk sac, withdrawn into the body during hatching, is still providing nutrition. However, ducklings often peck at each other, especially their eyes, in their curiosity. So by twenty-four hours, it is often a good idea to give them something to do. Introduce food by sprinkling crumbs on the floor, so that they are attracted by the movement and start to feed; they then move on to feeding from a container. Make sure the ducklings have already started using the water pot.

Starter Crumbs: The First Two Weeks

It is better to use a brand of starter crumbs rather than make up you own impromptu ration, so plan the food in advance. Try to obtain duck starter crumbs, free from coccidiostat, rather than chick rations, which will contain it. Ducklings kept in good conditions do not suffer from coccidiosis and it is possible that some coccidiostats (there are different types) cause death in ducklings. If no food is available, ground-up layers' pellets will do for a short time only. Failing this, grind up wheat and mix it with oatflakes and a little dried milk, hard-boiled crumbled egg yolk and brewers' yeast. This provides complete proteins and some vitamins, but move on to a balanced ration.

The crumbs are best fed ad lib in a heavy pot that cannot be turned over. It should be close to the water supply initially because the ducks will move rapidly to and fro as they alternately eat and drink. They usually manage dry crumbs very well this way but if there are problems in eating, moisten their food. If the crumbs are not fed ad lib, the ducklings will rush to eat at feeding time, stuff themselves full and nearly choke on dry food. When they drink, the food swells up inside them, making them gasp for breath. Avoid this situation by continuous feeding except for overnight, if using dull emitters. Always make sure the ducklings have sufficient water for the quantity of food they have available, i.e. the food will run out before the water does. After a few days, also separate the water container from the food to make the birds exercise themselves.

Starter crumbs contain a high percentage of protein, often increased by the addition of fishmeal. They also contain a higher percentages of minerals and vitamins than layers' rations. Chick starter crumbs may not contain sufficient vitamin B and D for ducklings, so do try to get duck starter food. Do not keep ducklings indoors longer than necessary; get them out into sunlight so that the vitamin D can be utilized. For vitamin B, sprinkle brewers' yeast on the food; the ducklings do like the flavour. These vitamins guard against rubbery beak (which may develop in ducks kept too long indoors) and weak legs. Purchase tubs of brewers' yeast from a chemist or obtain animal grade brewers' yeast from a farm or smallholder supply. Use in a ratio of one part yeast to forty parts of crumbs.

Growers Pellets

The ducklings should not be fed on the high-protein starter crumbs for long because the protein content is too high. The crumbs are also too fine for larger ducklings, and they waste them. As soon as their mouths are large enough they should be fed pelleted food in the form of growers' pellets suitable for ducklings (chick growers' will also contain coccidiostat). The pellets should be introduced as about 10 per cent of the food ration to start with and the proportion gradually increased until all of the ducks are coping with the pellet size. Call Ducks, for example, will have to stay on the smaller crumb size for longer because their beak is so much smaller than a Rouen's. Ducklings grow well on a balanced commercial ration but they also like, and benefit from, chopped green food such as shredded cauliflower and dandelion leaves and grass clippings. They should also have access to coarse sand and fine grit for the gizzard, especially if they are given fibrous material.

Large ducks are thought to require fairly high protein food to grow to as much as 12lb (5.4kg) in the Aylesbury and Rouen, and there was the idea that small ducks like Calls should be fed low-protein rations containing more milled wheat and less protein, in order to keep their size down. Whilst under-feeding certainly does hold back the growth of large ducks, Calls should not have their food limited to keep them small. Their size is partly an inherited factor, and late breeding of Calls is more effective in cutting their size down rather than under-feeding, which should not be practised.

As can be seen from the table below, layers' pellets do not have the same vitamin and mineral content as growers' pellets. If ducks are kept intensively and fed solely from bags, they should have the appropriate manufactured ration.

Rearing Batches of Mixed Breeds

Commonsense must be exercised if two or three breeds of ducks are hatched and reared together. If the breeds are roughly the same size, there will be no problems. If, however, larger ducks are hatched with Calls or Appleyard Miniatures, they will need to be separated after a few days. At first, having hatched two days earlier, the more active Calls sit on top of the larger ducklings. After a week's growth, the two sizes need separating.

Transferring to Out-Door Coops and Runs: Weeks Three to Six

Between two and three weeks, depending on the weather, the ducklings must be moved outdoors because they have grown so large, and they need daylight and green food for strong bones. A rat-proof coop and run is essential to provide protection from predators and the weather. Rats will take the

Duck starter and grower rations compared with a layers' ration

	Duck starter	Duck grower	Poultry Layer
Oil	3.75 per cent	4.25 per cent	3.5 per cent
Protein	18.00 per cent	15.00 per cent	16.00 per cent
Fibre	3.75 per cent	7.00 per cent	7.00 per cent
Ash	6.00 per cent	7.00 per cent	15.00 per cent
Methionine	0.35 per cent	0.25 per cent	0.35 per cent
Moisture	13.8 per cent	13.8 per cent	13.80 per cent
Vitamin A	13,500iu/kg	13,500iu/kg	8,000iu/kg
Vitamin D3	3,000iu/kg	3,000iu/kg	2,000iu/kg
Vitamin E	48iu/kg	48iu/kg	10iu/kg
Sodium selenite (selenium)	0.31mg/kg	0.31mg/kg	not specified
Copper sulphate copper)	25mg/kg	25mg/kg	not specified

The rations also contain calcium carbonate, di-calcium phosphate, sodium bicarbonate and salt – in the correct proportions

Fig 175 This drinker with a large rim is suitable for three- to four-week-old Call Ducks, as long as they have no eye problems. Their drinking water stays relatively clean because they cannot swim in it. Give them a bowl of water if they need a wash.

Fig 176 A useful coop and run for Call Ducks, which are small and have to be protected for several weeks before they can be released. The grass run is protected from the weather by a plywood door, which opens to allow the birds to be fed. The tongue and groove boarding protects them from wind. Polythene can be put over the coop if the weather is wet.

Fig 177 This larger coop and run is the best I have found for rearing waterfowl. The night compartment has a wired floor, which can be partially covered with a sack and straw for warmth. The mesh area allows droppings to fall through, and the ventilation keeps the bedding relatively dry. The run has plenty of room for drinkers and food and is covered over by the wire panel to exclude predators.

ducklings overnight if the coop is not sound. If the ducklings are not in a wired grass run they will be picked off by magpies and crows, as well as cats. If the weather is wet and windy, cover the wire of the run with polythene because this will raise the temperature of the run considerably, so that the birds are not huddled in the coop section all day for warmth.

Introduce the ducklings into the coop first onto their bedding of chopped straw. This can be put on top of a liner such as a paper sack, which will aid cleaning later when the sack and bedding can be removed in one piece. The birds usually find their way out to food and water via the pop-hole, but this needs checking. Also check that the ducklings know how to get back into the warm coop. It is sometimes necessary to pick them up and put them back the first time they have been out. Having slept in the coop one night, they then know what to do.

The pop-hole must be closed at night to exclude rats. Check that the coop has adequate ventilation for the birds. Many coops only have a couple of small holes, which may be insufficient; use your own judgement. The ducklings generate a lot of heat inside the coop, which is good because it keeps them warm. However, as the birds grow, the temperature may become excessive and the ducklings need to be let out early on sunny summer mornings to make sure that they do not over-heat.

Always keep a lawn of clean, cut grass for young birds. A field grazed by sheep is also suitable. The area should be free of adult ducks and their droppings because the ducklings need a good start in life, free from any bacteria and viruses that the adults may be carrying. It also gives the birds clean grass to pull and root around in to find insects. It is good practice with all livestock to rear it in as clean conditions as possible to avoid high mortality rates. The coop should be moved onto a clean patch of grass each day. This prevents the build-up of parasites such as coccidia, which take several days to develop to the infective stage. Ducks reared in dirty conditions can develop a low level of infection by coccidia, which damage the gut, which is then further damaged by bacterial infection, leaving necrotic tissue. Such birds do not grow well and eventually become emaciated and die.

Always make sure, both for adult ducks and young stock, that the ground is free from hazards such as string, wire, staples, glass and nylon bristles from brooms. It is surprising what worms and moles will bring to the surface for the ducks to pick up.

Ducks are adept at escaping from runs because they have their heads down, rooting in the grass. Make sure that any potential escape holes, where there are dips in the ground, are blocked up by bricks or wood. Escapees can get lost, and are soon picked up by crows.

In the run, there is scope for larger containers for food and water. A larger drinking fountain can be used, or washing-up bowls for taller birds like Indian Runners. Make sure that the ducklings can get out of a bowl by putting a brick inside it as a stepping stone. The water may get fouled with droppings but on clean ground this does not seem to cause problems with rearing. The water container should be placed at the far end of the run, away from the dry bedding of the coop in order to keep it as dry as possible.

Avoiding Problems

Do not keep the ducklings in the coops longer than necessary. With the larger breeds, they can be out by four weeks if the area is free from predators such as crows. It is particularly important to get Indian Runners out to use their legs and develop strong bone. Calls need to stay in runs for longer simply because they are smaller and need the protection. They are not adversely affected by being confined.

Always turn out the birds onto clean ground, or give them their liberty in the area where their own coop has been moved around. This way, they are not introduced to new germs. If they have to follow on to the patch of the preceding hatch, try to rest the ground for a week or two to allow the droppings to be washed into the ground by rain.

Feather Picking

Ducklings in runs get bored and may start to pull lumps of fluff or blood quills from other ducklings. If certain birds are losing fluff (on the back or, particularly, the quills full of blood on the wings) and looking edgy, watch the ducklings to see what is happening. When the culprit is identified, remove it and move it up an age group. The larger birds are better able to cope, and may be past the blood quill stage. The worst breed for feather picking is the Rouen, where nutrient demand may be the problem, but it does occur in Cayugas and Calls. It is said to result from protein and salt deficiency. Higher protein food can be tried with Rouens because they grow very rapidly, and old duck handbooks did recommend meat scraps in duck food. The salt content of the foods should be balanced, but a complete mineral supplement could be given for a short time. Do not add your own salt (sodium chloride). With small ducks like Calls, the problem is usually boredom and habit.

Wing Problems

Ducks that grow too quickly carry blood quills, which are too heavy for their wings to support. This results in:

- dropped wing, where the quills are too heavy to carry in the normal wing position – the duck may pick the wing up later and be all right;
- rough wing, where the primaries are not folded neatly in their normal position but sit slightly turned out – this spoils a bird for showing;
- slipped or oar wing, where the final joint is turned outwards – this only occurs in domestic birds that have access to large amounts of food; it would be lethal in a wild population because such a bird cannot fly.

Avoid these problems by under-feeding pellets between weeks five to eight (during the growth of the blood quills and the sprouting of the feathers). Start to introduce wheat in the diet at five weeks, increasing the amount to 50 per cent, if necessary, by week seven. Also underfeed the birds at this stage and check that the pellet ration is not too high in protein – 15 per cent rather than 16–18 per cent. Once the feathers are fully formed and set, more pellets can be fed to get size in the larger breeds. It is essential to provide access to grit and sand for the ducks when they are feeding on hard grains, so that the food is milled up in the gizzard.

Chilling

Even at six weeks old, the ducklings still have fluff on their backs; their underparts develop protective feathers first. If the weather is very wet, the ducklings get chilled with continuously wet backs and certainly need housing in a dry, warm shed overnight. They can only be left outdoors in a fox-proof pen overnight when they have a complete covering of feathers.

Sunstroke

Ducklings are particularly prone to sunstroke, and Pekins are the most susceptible. The birds are listless and not interested in eating or drinking. Some of them may die. Provide shade by erecting secure plywood, or cut cypress branches and lean these in wigwams to make shade. For the long term, plant shrubs. Make sure plenty of clean water is available.

Swimming and Drowning

Although ducklings will jump into their water containers in their grass runs at three to five weeks, swimming is neither necessary nor recommended whilst they are fluffy. Domestic ducks reared without a mother are not as well oiled as Wild Mallard. If ponds are available, make sure that they are shallow and the ducks can get out of them easily. The sides should not be slippery because birds become water-logged and will drown if they cannot get out. If the pond water is dirty and stagnant, it is better if birds do not have access to it. Any bird that is wet and chilled (at any age) should be blotted dry and put under a heat lamp (supervised) to revive. Birds that seem dead with cold can revive.

Disease: Botulism

As well as avoiding dirty ponds, do not allow the ducklings access to boggy areas in summer. Rotting

Fig 178 This Rouen drake, which turned out to be the largest, grew very rapidly. The primaries of one wing turned outwards, and should have been taped into the correct position earlier.

organic material allows botulism to flourish in the heat. Ducks pick up the disease as they search for worms.

Disease: Coccidiosis
Avoid this by folding ducklings on clean ground each day when they are in coops. Cider vinegar, stocked in large quantities in agricultural stores, has long been used for avoiding coccidiosis in larger animals. It is now also recommended for poultry and ducks for controlling bacteria, worms and coccidia.

Disease: Enteritis
There are different forms of enteritis (viral and bacterial). The bacterial form is most likely to arrive in a hot spell in summer and affect ducklings. The smallest breeds – Calls and Appleyard Miniatures – are affected first. They become inactive and die. Prompt treatment with antibiotic will save most of the birds if the disease is the bacterial form. Move them onto clean ground; rest and lime the vacated area. If you buy new birds, particularly in summer, quarantine them from the home flock for two weeks.

Worms
Well-fed ducks do not seem to suffer from worms. The only one that occasionally bothers them is trachea worm (gapes). Where ducks or chickens have been reared on the same ground for years, there is a build-up of this parasite. Adult ducks are rarely bothered by it but young ducks, at about six to twelve weeks, may cough. This is especially so if they are in damp pasture with lots of earthworms, which are a host of the parasite. It is good practice to treat the ducks with a wormer in the food or water *(see* p.182).

Sexing Ducklings

Whilst the ducklings are growing, there are many indications of their sex to the practised eye and ear. Even by two weeks, some females are starting to give a deeper voice than the usual peep.

Voice
Usually, the ducklings have to be four to five weeks old before one can be certain of the sex of all of them by voice. On picking them up, drakes can give quite a deep, hoarse noise like a quack if they are afraid. It is best to listen to them relaxed, at close quarters, to decide which are the females.

Colour
In coloured breeds, there are several indications of the sex of the birds as early as five to six weeks. The diagnostic feather area is the rump. Drakes have something like the characteristic adult male colour, e.g. glossy green-black in Rouen, whereas females have the brown ground colour showing. The contrast in rump colour is a useful diagnostic feature in many breeds such as Fawn-and-White Runners, Appleyards, Abacots and coloured Calls.

The colour of the beak also indicates the sex, particularly in breeds where the adult male has a green beak. This colour will begin to show by weeks four to five.

Sex Curls
White birds do not show auto-sexing colour characteristics, the curl on the rump of all drakes at sixteen weeks and over is the only diagnostic feather characteristic.

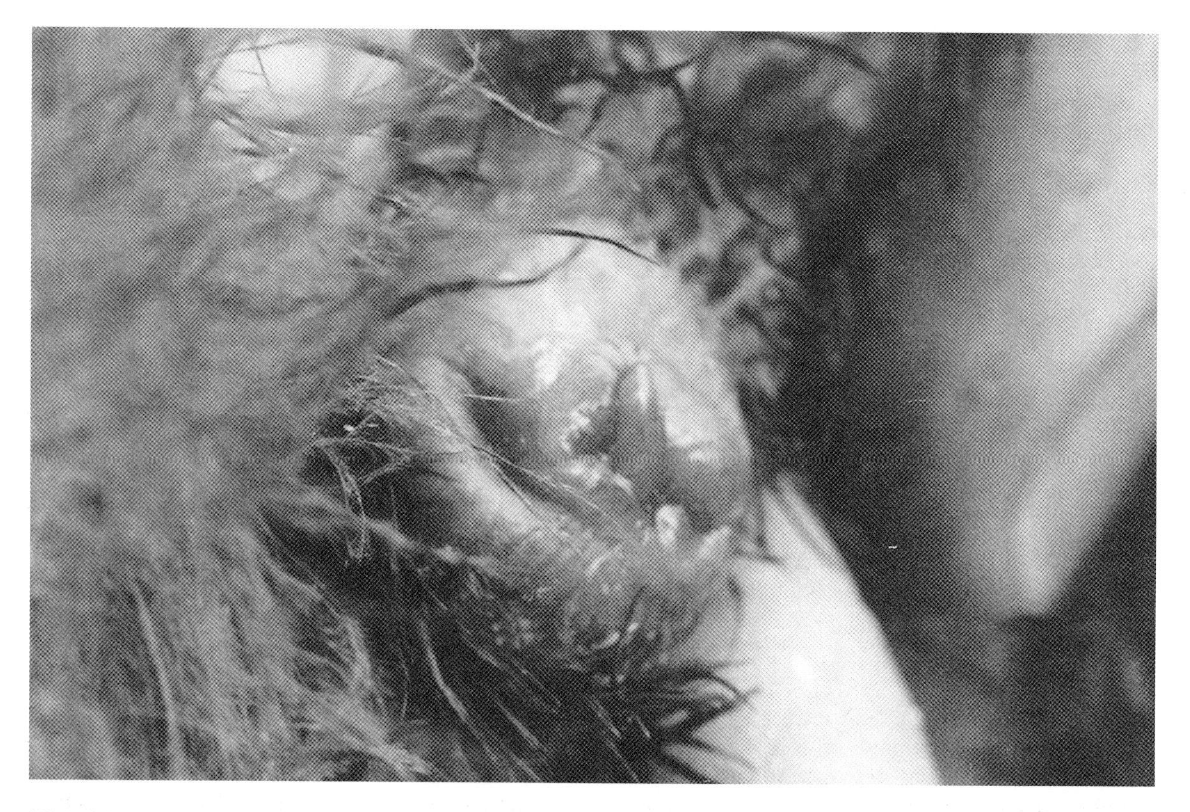

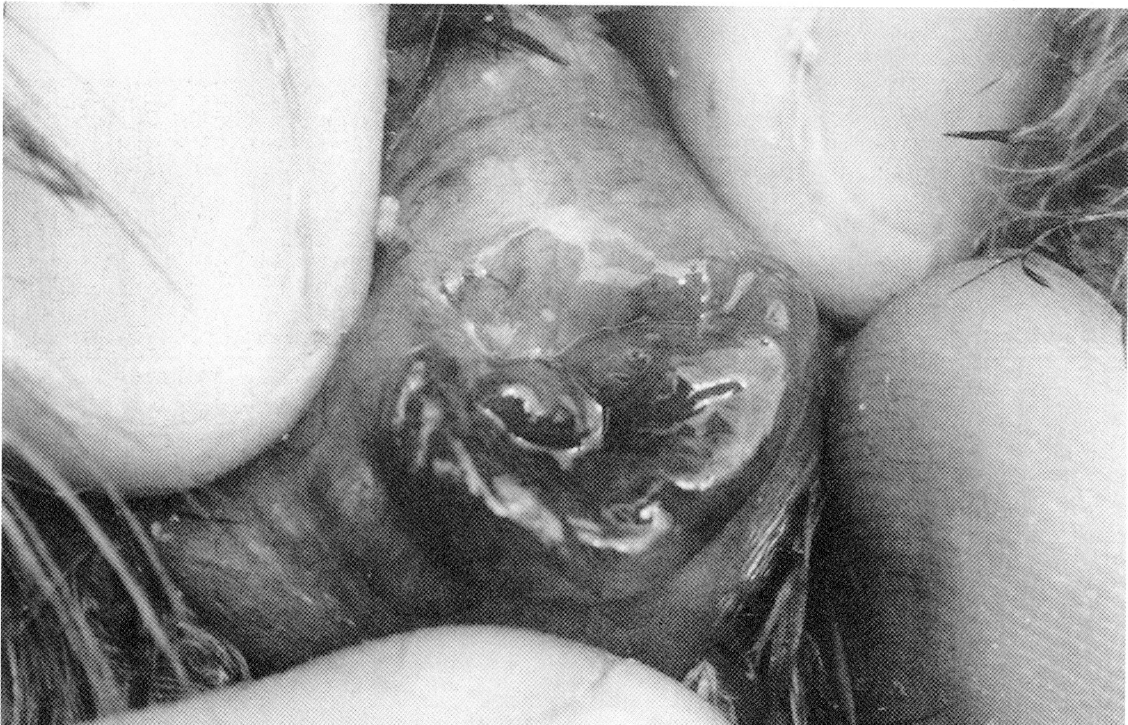

Figs 179 and 180 Vent-sexing ducklings. (top) A small penis is exposed in this three-week-old male bird. (bottom) The cloaca of a female of the same age.

Size

Drakes are usually bigger than the ducks, for the same strain, e.g. Runner drakes are taller than ducks.

Vent Sexing

It is often convenient to know the sex of the ducklings earlier than four to six weeks if they are not to be reared to the adult stage for breeding and exhibition. Surplus drakes can be sold for rearing and fattening. If only females for laying are wanted, then the drakes can be culled, as in commercial hatcheries. However, one has to be practised at vent sexing for this to work well with young ducklings. They are very fragile, and there is the possibility of injuring the bird. If sexing has not been practised before, get someone with experience to show how the birds should be handled.

Practise first on a three- to four-week-old duckling that you think is a drake. Use a well-lit place (outdoors or with a direct light). Hold the bird across your knee, on its back. Holding it down with your arm, use both hands to push back the tail of the bird so that the vent is exposed. Part the fluff and apply pressure with two fingers to the sides of the vent. Then, using the thumbs as well, apply pressure to the top and bottom to evert the vent. The penis should pop out as a spike, about 3mm long. Compare this bird with one thought to be a duck.

Only when you have practised this technique should younger ducklings be used. The penis of a drake a few days old is very small and difficult to see. There are usually more drakes than ducks, though this can vary from year to year.

Separating Young Ducks and Drakes

For most of the summer growing period, the young growing ducklings can be kept together. As the birds mature, however, the males and females may need to be separated. Rouen take a long time to mature and the drakes are not generally interested in mating until spring. White Runner drakes, on the other hand, are especially precocious and will be mating with females in the autumn. Apart from stopping the females getting pulled about, the drakes need to be looked after too. Enthusiastic, frequent mating can result in prolapse of the penis in juveniles. They do not always manage to recover from this and have to be put down. It is therefore often best to separate the birds into a pen of ducks and a pen of drakes for easier management. Give the ducks the better conditions; they are more valuable and are coming into lay. If the weather is cold, they are better housed at night to keep them warm while they are laying. It also allows easier collection of the eggs.

Marking Ducks and Ducklings

For a pedigree breeding programme, ducks need to be marked. It is essential to know which birds are the parents and offspring, and to mark any new stock birds brought in. Feather clippings are useful but only last part of the year, until the next moult.

Adult ducks are generally marked by using plastic spiral leg rings. These can be quite difficult to get on when the ring is of a small diameter. Warm up spiral rings before fitting them because this makes them more flexible and easier to manage. For the smaller breeds, such as Calls, spirals cannot be used. Flat plastic coiled bands can be fitted more easily. The smaller sizes tend to stay on better than the larger ones; the ducks tend to catch them in something and pull them off. A new, better type of leg ring simply pulls open and closes automatically because of the reinforced plastic at each side. All of these kind of temporary markers must be removed for showing.

Permanent markers can be achieved by:

- Wing tags: these are fitted through a neat hole punched in the skin only, where the wing joins the body. Occasionally these are partially pulled out and cause a local infection. They are generally trouble-free.

Fig 181 Different kinds of leg-rings for marking ducks, available from poultry suppliers. a. Spiral plastic ring: not easy to get on and off. Use for a permanent marker. b. The larger sizes of spiral rings are more flexible and so easier to get on and off (suitable for Pekin and Rouen). c. Reinforced plastic, which pulls open and closes automatically. Not as widely available as the above two types. Useful for Call Ducks. d. Flat plastic coil: easy to get on and off, but the ducks can also pull then off by catching them in wire. The smaller Call Duck sizes stay on better than the larger rings.

- Tattooing: inked markers are fitted into the kit (bought from agricultural stores). Again, only the skin is squeezed where the wing joins the body.
- Closed leg rings: these must be fitted when the ducklings are three to four weeks old, when the foot will pass through the ring. The rings are made in several different sizes for the different breeds. As the duckling grows, the ring no longer passes over the foot. They may have to be refitted a couple of times. There are different suppliers and prices, e.g. The Poultry Club and Aviornis. The rings are numbered, are a different colour for each year and are made of metal. This is a disadvantage if birds live out in a fox-proof pen. In cold weather the metal ring may freeze onto a surface. Also, the rings can get caught and imprison the duck.
- Split metal leg rings, which are numbered and closed onto the leg of the bird using special pliers (supplied by Aviornis).

These permanent marks are not removed for shows.

Humane dispatch of birds

Breeding a number of birds inevitably means that the producer must accept the fact that there will be birds that need to be put down if they are unhealthy or are drakes that should be slaughtered for the table because they are surplus. If the birds are specifically bred for the table and are to be slaughtered in large numbers, then it is essential that, at the very minimum, Ministry regulations are followed and that the latest guidelines are obtained from MAFF. If only a few birds are kept as pets, then any aged or sick birds may be best put down by a vet.

Birds for Table Purposes

Withdrawing Food
Before slaughter, food should be withdrawn for at least four hours so that the gullet is empty. This period can be the normal overnight period of up to twelve hours. The ducks should not be left for prolonged periods without food or water.

Handling Birds Before Slaughter
Birds should be handled quietly before slaughter. If they are in a large group, they should be split into smaller groups in a shed, e.g. by using wire partitions to separate them. This avoids the birds panicking and rushing into a corner. Large groups of birds will pile up and smother each other, and scratch each other with their claws. To pick them up individually, herd the ducks in groups of about eight. There is no need to handle them before this stage if moveable weldmesh partitions are used to direct the birds into a convenient area.

Ducks should not be caught by the legs; they are easily damaged. Always catch the bird using both hands either side of the body. The duck can also be restrained by the wing or neck but the body weight must be supported if it is lifted.

Methods of Slaughter
A poultry slaughterman's licence is needed for commercial slaughter (thought not if neck dislocation or decapitation is used on the farm where the bird was reared). Contact the local MAFF Animal Health Office on how to obtain a certificate of competence. A licence is not needed for private slaughter, but humane methods must be followed. It is an offence to cause any animal avoidable suffering. If you have not practised slaughter methods before, it is essential that you read up-to-date MAFF regulations (The Welfare of Animals [Slaughter or Killing] Regulations 1995). Recommendations in old text books may well be illegal now. Contact a local Agricultural College and the Humane Slaughter Association to see if a course is available, or the Poultry Welfare Officer Training Course at Bristol University. Otherwise seek advice from someone who knows up-to-date humane slaughter methods and regulations. You should take advantage of someone else's experience.

Legal methods of slaughter include the following. Note that the Humane Slaughter Association also recommends stunning prior to all of these methods, particularly if large quantities of birds are to be handled. This should be done with an electrical stunner.

- Electrical stunning that induces a cardiac arrest has been shown to be the preferred method on welfare grounds. This should be followed by bleeding (severing of the both carotid arteries and both jugular veins). This method is recommended for larger numbers of birds. In all cases when electrical stunning is applied, birds should be restrained to control wing flapping and to ensure correct application of the electrodes. Make sure the equipment used is maintained and used correctly.
- Decapitation with a heavy axe on a wooden block. However, brain activity may remain for up to thirty seconds, and can be longer. Electrical stunning first is recommended.
- Neck dislocation:
  - by hand: this is easier to carry out on hens than large ducks;
  - using a stick to dislocate the neck: this is better

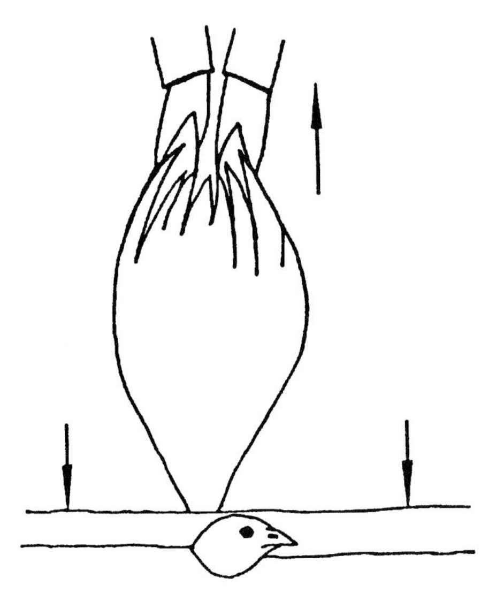

Fig 182 Method of slaughter: dislocation using a heavy stick. (By permission of the Humane Slaughter Association, Practical Slaughter of Poultry: a guide for the small producer; see Appendix for details)

Fig 183 The 'Cash' poultry killer.

for larger or longer-necked birds. The bird is held by the legs and wing tips, with the head and neck on the ground. A heavy stick or bar is placed across the neck, behind the head, and firm pressure applied to the bar each side, only when one is ready to pull the bird's body upwards. The neck dislocation is achieved by using one's weight across a broom handle across the neck. This pressure dislocates the vertebrae and severs the spinal cord. With neck dislocation, always check that the bird is dead by feeling the neck for a finger's width gap in the vertebrae.

Read the Humane Slaughter Association's guidelines, which give more details and also obtain their most recent literature on a poultry killing device that delivers a blow to the bird's head killing it instantly. The 'Cash' Poultry Killer has been designed to kill chickens and turkeys at all stages of growth from day-old to adult, but work is currently in progress to assess its effectiveness on ducks. The results are therefore awaited.

The percussive device comes in two models, both designed to apply a powerful blow to the bird's head, killing it outright. However, if the device is used for commercial purposes, it must be followed by bleeding or neck dislocation to satisfy current legislation. There are two versions currently available:

- the first is a pneumatic device, for use as back-up to the electrical stunner and killer in the processing plant – this has to be used at a fixed place because it is powered by a flexible airline.
- the second is a mobile cartridge-powered device for killing casualty birds on-farm.

The equipment is manufactured and marketed by Accles and Shelvoke Ltd, PO Box 705, IMI Industrial Estate, Brookvale Road, Witton, Birmingham, B6 7UT (Tel. 0121 344 3155).

Young Ducklings

Occasionally, young ducklings are hatched with an obvious deformity that is, or is going to be, a real problem for the bird's welfare. Such ducklings should be put down as soon as possible. The recommended way for dealing with these on a small scale it to dislocate the neck. This can be done by pressing the duckling's neck hard against a sharp edge, such as a right-angled table top, to dislocate the vertebrae. Neck-stretching techniques are preferred to the use of Semark pliers, as these crush the neck and do not directly damage the brain.

Appendix: Keeping Ducks Healthy – Preventative Care

Disease prevention is far more important than attempts at disease treatment and control. Ducks are generally healthy birds and do not suffer from many ailments. Following this summary of good practice will reduce disease to a minimum.

- Be observant. Good stockmanship is an art that is acquired through interest in the birds. Watch stock closely for signs of unusual behaviour. Birds are very good at concealing the fact that they are ill, until it is too late. If any bird is behaving differently from usual, pick it up to have a closer look at its eyes, nostrils and vent, and to check its weight. When birds have to be collected and driven into a shed each night, rather than being left out in a fox-proof pen, symptoms of illness are more likely to be recognized.
- Noticing signs of illness early saves birds. Medication is likely to be more effective. Also the sick bird can be isolated from the flock, reducing chances of further infections.
- Provide clean water. This allows birds to wash their sinuses, to clean their feathers and preen effectively.
- Make sure that all food is stored correctly and is free from moulds. Do not leave food in dishes to go stale. If possible, buy pellets manufactured for ducks, not hens.
- Provide grit for the gizzard.
- Keep the ducks' night quarters dry and well ventilated.
- Remove all foreign objects that could cause injury, such as string, plastic, fragments of wire, and so on. Check for hooks and nylon line in fishing areas.
- Check birds for external parasites, such as mites, lice and even leeches.
- Worm birds if they are thin; otherwise worm routinely once or twice a year with Flubenvet (flubendezole), which treats all types of worms.
- Make sure the birds do not contact poisonous materials such as rat poison, creosote, lead shot, weed killers, nitrate fertilizer (in granules or as run-off from recently treated fields).
- Do not allow birds access to rotting animal and vegetable waste, septic tank waste or farm slurry.
- Provide shade and plenty of clean water in a heat wave.
- Protect from frostbite in severe winter weather.
- Handle birds quietly; do not feed immediately before catching.
- Rear young birds on clean ground that has been limed and rested. Rear them separately from adults to avoid high concentrations of pathogens before young birds have had time to acquire immunity.
- Remove all manure from the premises if possible; spread on land used by other types of animals.
- In commercial production especially, where large numbers are involved, scrupulously clean sheds, yards, feeding and drinking equipment.

When to Consult the Vet

If you only have a few ducks, and birds are quarantined when new, there are unlikely to be many problems. The commonest ailments are infected leg joints and egg-binding in females during the laying season. Other problems do occur and a consultation with the vet is useful to find out more about treatments that could then be administered by yourself in the future.

It is often quite impossible to tell why a bird has died unless your vet sends off a tissue sample to the local Ministry of Agriculture Pathology Laboratory. This is a time-consuming and expensive process and, unless you have several birds dying inexplicably, not worth doing. By the time you have the result it is either too late, or medication resolved the problem anyway. It is no use waiting ten days for a result! In contrast, your local vet can usually do a post-mortem to check for the likelihood of TB, liver infection or coccidiosis. The advantage of having a laboratory report, however, is that the correct treatment can be prescribed if the symptoms recur in other birds. Disease in birds must be caught early if treatment is to be successful.

Advice will be needed on choosing the correct medication. Coccidiosis is not caused by bacteria, for example, and therefore does not respond to antibiotic treatment such as penicillin. Diseases caused by bacteria do not always respond to a particular antibiotic; the correct one is needed for specific conditions. The antibiotics cited in the text are the ones often used in treating ducks but the quantity and the type should always be checked with a vet before use. Note that some of the antibiotics used generally for bird treatment have not been specifically developed or tested for birds. For some coccidiostats the dosages may be for species of birds other than Mallard. Wormers such as levamisole and Panacur have not been clinically tested for birds either. Dosages in these cases are based on practical experience, but adverse effects may occur and should always be watched for.

Ailments and Diseases

Aspergillosis

Spores of various fungi proliferate in damp, mouldy bedding, particularly hay. Only buy straw that smells sweet and has been dry-baled. The spores cause 'farmer's lung' in humans and cause respiratory distress in birds too. The fungus grows inside the respiratory system and gives the symptoms of pneumonia. There are fungal treatments now available (similar to those used for humans in fungal infections) but they would not be economic to use for ducklings, which sometimes develop laboured breath and lung congestion (which could be aspergillosis or pneumonia). Avoid problems by removing damp litter promptly from the rearing area.

Adult ducks are unlikely to suffer from aspergillosis if rough whitewood sawdust is used for bedding. Good quality straw plus shavings should be used in the breeding season for nests in a shed, and bark peelings for damper places outdoors. Nesting material must be clean otherwise *aspergillae* can be transferred to the incubator and affect the ducklings (Pearce, 1979).

As well as using mould-free bedding, check that all food is stored in dry, cool conditions. Do not accept pellets that are turning blue-green; return them to the manufacturer. Never feed mouldy bread.

Avian Tuberculosis

TB can be carried by wild birds (ADAS Leaflet 1, 1980). The disease takes a long time to develop, and the bird will become listless, thin and die. A post-mortem of such a case will quickly reveal infected joints and an enlarged liver, studded with small white or yellow nodules. All associated birds should be tested (very much as a human TB test). The pin-prick test to the skin should be applied to the side of the bill at the fleshy base, where it is attached to the head. Do not allow the feet to be used; the birds can go lame. Affected birds react by a swelling at the test point and must be culled and burned. Further testing is necessary at one to two month intervals until the flock is clear. The birds should preferably be moved onto clean ground and the housing disinfected.

Other liver infections such as ulcers caused by *E. coli* can superficially mimic TB, and are more common.

Botulism

Botulism results from poisoning. Toxins are ingested from boggy areas with rotting vegetation or animal matter, particularly in warm spells in summer. The toxins are caused by a species of *Clostridia*. An affected bird rests its head on the ground because of paralysis of the neck, hence the term 'limberneck'.

Make sure that birds do not have access to boggy areas, shallow stagnant pools or decaying animal matter. There is no treatment for the disease (Curtis, 1996) other than removing the source of the infection, providing clean drinking water to flush out the system and hoping for the best. Mild cases will recover. Antibiotics help stop further deterioration of tissue.

Bumblefoot

A colloquial term referring to a swelling on the foot caused by bacteria (*Staphylococcus aureas*) entering through a bruise or injury. Antibiotics may help initially, but there is no really effective treatment other than surgery to remove the affected tissue. Old, heavy ducks such as Rouen may develop the condition (*see* page 114) but do not seem troubled by it as long as they are on soft ground and have access to plenty of clean swimming water.

Coccidiosis

Coccidia are species specific (ADAS Leaflet 16, 1979). These protozoan parasites live in the lining of the intestine and damage it. Birds are unable to absorb nutrients from food, so that an infected bird becomes thin and looks miserable. Ducklings will fail to grow. Continued damage results in further infection and death. Symptoms are lethargy and loss of weight, and particularly blood in the droppings. Look at droppings passed in the morning, just before the birds are released from the shed; any blood can then be seen. A vet can gauge the level of infection from a sample and the birds treated with an anticoccidial. The coccidiostat added to some poultry growers' foods is not a treatment. Ducks are less affected by the disease than poultry, but it can be problem if they are on dirty ground covered in droppings, particularly in the summer heat in damp weather. This disease may, however, occur at any time of the year and affect adults.

Coccidia oocysts are highly resistant and given warmth and moisture will proliferate. Oocysts are present on all poultry farms and even the healthiest birds are likely to be affected with a few parasites. These birds become resistant to further attack. Fatal disease follows in birds that have failed to acquire resistance, which follows light infection. Oocysts take two days to 'ripen' after they have passed through the gut so that moving coops onto clean ground each day keeps down infection in young ducklings. The addition of cider vinegar to drinking water is said to combat coccidia and worms; use 25ml of vinegar to one litre of water, in a plastic container (Earle, 1998; *Fancy Fowl*, October 1999).

Crooked Toes

These can be a genetic defect, or caused by incorrect incubation temperature.

Crop Binding and Sour Crop

Ducks do not have a crop to store food, unlike the chicken. Food is stored in the gullet and proventriculus in a similar way to the crop, hence the term 'crop binding' when things go wrong. Avoid problems by making sure that the birds have access to grit for grinding food in the gizzard. Also, keep grass short so that it is tender; long, stringy grass will cause problems.

Crop binding and 'sour crop', when the proventriculus begins to swell because the gizzard is malfunctioning, is rare in ducks. It is more likely to occur in Toulouse geese and chickens. You can try to treat this condition with massage to break up the lump whilst holding the bird out or its neck downwards. The bird will have difficulty in breathing in this position, so do not persist. If you are successful, also treat the bird with antibiotic in the water. The bird can be operated upon by the vet, to clean out the impacted debris, but the condition usually results in it having to be put down. An old remedy (e.g. in Cook, 1899) was to use magnesium sulphate crystals (Epsom salts) to increase gut motility, and this is still recommended.

Duck Virus Hepatitis

Ducklings stop eating, though they continue to drink. Their movements become sluggish. The birds fall on their sides and paddle their legs ineffectually. A few birds recover. Because this disease is viral there is no immediate treatment. The disease is avoided by using clear or running water rather than stagnant pools. Where the disease and water supply is a problem, it is best to vaccinate the birds and improve the water supply.

Enteritis

Enteritis is a general term for inflammation of the gut. It can be caused by several bacteria or viruses.

Bacteria

This is most likely to strike young ducklings in a hot spell in summer. It is probably introduced by wild birds carrying the infection. Small breeds such as Appleyard Miniatures are affected first. The ducklings are listless and do not want to eat, but will drink. They may die within twenty-four hours of symptoms being noticed. Prompt treatment with an antibiotic such as Terramycin (oxytetracycline) or Aureomycin (chlortetracycline) prescribed by the vet in the drinking water will save most of the birds. Keep the birds and the medicine in the shade and remove calcium, i.e. grit, temporarily (this may inactivate some drugs). Soluble powder medicines lose their strength rapidly after being made up, and should be replaced at least once a day. Once treatment has started, move the ducks onto clean ground; treat the vacated area with quicklime.

Viral

Known as 'duck plague', this was first diagnosed in the UK in 1972. It was believed to have been introduced by migratory birds from Europe. All of the *Anseriformes* are susceptible and losses are high. Signs of the disease are listless birds with drooping wings, diarrhoea causing a matting of the vent feathers, and a dirty appearance of the head from increased secretions from the eyes and nostrils.

Depending on the stage of the disease, haemorrhages or necrotic tissue is found from the oesophagus right through to the cloaca. Hall (1973) described the disease in detail and considered that the extensive occurrence of this necrotic membrane provides strong evidence for the diagnosis of plague. Minute haemorrhages of the heart muscle, the pancreas gland and the body membrane were also fairly commonly found.

Infection from migrating wildfowl, spread by droppings in the water, was the most likely cause of infection. Unfortunately there is no antibiotic treatment available because the disease is viral. It is now possible to inoculate the ducks. Contact Intervet.

Eye Problems

Pekins are more prone than most breeds to develop eye infections. This is because their abundant feathers tend to turn into the eye, but they also seem to be more sensitive. The condition is usually worst in muddy conditions, suggesting that mud does irritate the eye. There are four things that can be tried to cure the condition:

- make sure the duck is free from northern mite – scratching can cause eye irritation;
- keep the ducks in mud-free conditions if possible;
- before resorting to antibiotic eye cream, try confining the bird in clean conditions for a few days with access only to water from a bucket to which has been added a few drops of Milton – this frequently cures the problem;
- breed from Pekins which have good eyes – do not use birds that have recurrent problems.

Sticky Eye

Ducks develop a sticky discharge from the eye if they have insufficient water for bathing. Runners are particularly prone to this. Make sure the ducks have more clean water. Feed wheat under water to make them bathe their eyes.

White Eye

A white cast may develop over the pupil and annoy the duck because it cannot see. Ducks often scratch at their eye and can make themselves bleed. This condition may be similar to New Forest disease, which affects the eye in cattle, particularly in summer. Treat with an antibiotic cream. This sometimes alleviates the problem. The white patch can also disappear on its own in time.

Eggs

Soft-shelled – *see* Reproductive System

Feather Picking

This occasionally happens with growing ducklings in the age range of four to eight weeks when feathers (especially the flight feathers) are sprouting and contain blood. Always remove the offending bird or birds and move them up an age group, i.e. the hatching group two weeks older, to stop the habit. The habit may be caused by boredom, protein demand or by mineral deficiency. Sodium deficiency has been cited, but this mineral is added to pelleted diets. Extra sodium as common salt should not be added to the diet.

If the vice is a problem with several birds it might be worth trying a complete mineral supplement and a higher animal-protein food. More space, green food and plenty to do does not always stop young Rouen in this habit, however, until they have completed their first set of feathers.

Fig 184 Keep Pekins is good conditions. They get dirty feathers and sticky eyes if access to water is poor.

Gapes
See under Worms

Lice
These live on feather and skin scale and can become a problem if present in large numbers. Fit ducks remove them by preening.

Maggots
During spells of hot, damp summer weather when flies proliferate, eggs may be laid around the vent of a duck. Birds with access to plenty of bathing water will not suffer because they keep themselves clean. The flies are more likely to settle on a bird that is already ill or injured. Pick the maggots off, use an ointment if the skin is raw and treat the affected area with fly spray.

Mineral Deficiencies
As with vitamin deficiencies, these are unlikely to occur if birds get a mixed diet with some greens. Calcium is added to poultry pellets in the correct amount for growers, maintenance and layers. Extra calcium should always be made available in the breeding season in the form of limestone chips and oyster shell so that the females can help themselves to the amount they want for egg formation. Excess calcium is not necessary for ducklings and the calcium content of layers' pellets is higher than they need and can be harmful as it will then increase their demand for phosphorus.

Other minerals such as magnesium, manganese, copper, iron, iodine, potassium, sodium, selenium, zinc, etc., are needed in trace amounts and the feed label on the bags of pellets will indicate that these have been added. Runner ducklings are more likely than most to suffer with wobbly legs and benefit from a mineral supplement, plus getting them out on grass to exercise their legs and eat greens early (*see* Chapter 8).

Mites: Northern Mite and Red Mite

Northern Mite
This is a red, blood-sucking mite that lives on the bird. In waterfowl, it seems only to infest the head and neck. It causes the bird to scratch and may be a contributory factor in eye-foaming in Pekins. It is difficult to keep birds completely free of the parasite as it is transmitted quickly from bird to bird. The parasite is difficult to see on coloured birds. It is more easily discovered on white birds if the feathers are parted on the crown of the head. If birds are scratching, they should be treated with a pyrethrum powder by ruffling up the feathers when the bird is dry, preferably when it is shut in for the night. The process should be repeated seven days later as pyrethrum is a contact killer and is not persistent. Other insecticides, which are non-toxic to the larger organisms, are now available and quite effective, e.g. Barrier dusting powder, which contains herbal concentrates and essential oils and kills red and northern mite, and Blast Off (available from The Bird Care Company).

Red Mite
This commonly infests chicken houses but can get in the duck house. If the birds are scratching for no apparent reason, try spraying the shed with insecticide instead. Duramitex is often recommended, but it is a very persistent organo-phosphorus chemical. Try a pyrethrum spray on the woodwork first, and repeat seven days later to kill emerging nymphs. This infestation is rare with waterfowl. The red mites live in the cracks in the shed and are not permanently on the bird; they can live for months without food. They look virtually the same as northern mite, i.e. they are red when they have eaten. If they have not eaten for weeks they are grey.

Newcastle Disease
A highly infectious viral disease causing respiratory problems, diarrhoea and high mortality. Occasionally visits us from the continent, e.g. in 1996–97, probably carried by migratory starlings. It seemed to affect large, commercial establishments rather than people with a few birds. When a case in reported, zones of exclusion are drawn, and poultry movement halted. This is a notifiable disease in the UK with compulsory slaughter for highly pathogenic strains (Curtis, 1996). Chickens and turkeys are much more susceptible than waterfowl, which are reputed to carry the disease but not suffer from it. A vaccine is available.

Pasteurellosis (Fowl Cholera)
The birds may have watery green droppings, difficulty in breathing, discharge from the mouth and nostrils and are unable to get up. The bacterium *Pasteurella multocida* causes respiratory infection and is spread by nasal exudate, as well as in the droppings, because enteritis is also associated with the condition (Laing, 1999). There is a loss of appetite, increased thirst and a high body temperature. Birds lose control of their neck muscles in the later stages.

Isolate sick birds. Treat with antibiotic such as Terramycin prescribed by the vet; if caught very early you may save the duck. If the bird will not drink, inject the bird instead with a suitable antibiotic. Also use a syringe (no needle) to make sure that the bird gets liquid down the throat and does not get dehydrated. Antibiotic can also be administered orally. Oral administration is only suitable with larger birds, unless a tube syringe is inserted down the throat. With smaller birds, liquids may accidentally enter the trachea and lungs.

This disease can be spread by wild birds, rats and mice and may occur at any time of the year, though it occurs most often in autumn and winter. Laing considers that the environment remains infected and healthy birds may remain as carriers, so improving hygiene after the occurrence of the disease is important. If possible, burn all affected carcasses. Otherwise bury them deeply. Lime the affected ground and rest it.

Pseudotuberculosis (*Yersinia*)

In dirty conditions, birds can develop internal white growths in the liver that resemble TB. Infected birds are thin and infectious. As with many diseases, this cannot be readily diagnosed except by post-mortem. Antibiotics can help. Uncommon, but can affect humans (Laing, 1999).

Reproductive System Problems

Calcium Deficiency and Soft-Shelled Eggs

Sufficient calcium must be provided for shell formation, so laying ducks should be fed layers' or breeders' pellets with the correct amount of the mineral. Calcium is also obtained from the limestone chips and oyster shell in mixed poultry grit, and from grass where pasture is at the ideal pH of 6.5–7. In addition to the calcium, vitamin D from the pelleted food or greens and exposure to sunlight is necessary for its absorption. If there is insufficient calcium in the diet, the duck's supply of calcium in her own skeleton is depleted and she will pass soft-shelled eggs, i.e. eggs that have a membrane but no shell. These eggs are sometimes laid normally but the situation can lead to oviduct problems, i.e. difficulty in passing the egg.

Calcium deficiency exacerbates the problem in two ways:

- there is no hard shell for the muscle contractions of the oviduct to push against;
- calcium is also needed for muscle contractions; inadequate contractions will therefore hinder the passage of the egg, i.e. cause egg-binding.

Egg-Binding

Birds may sometimes have extreme difficulty in passing an egg. If this is low down in the oviduct and the egg (with or without shell) can be seen as the bird strains to pass it, the duck can be helped. Bring the bird into a warm environment to speed up the bird's metabolism and relax the muscles. After the bird has warmed up, squirt olive oil into the vent with a syringe (no needle) and try to aid the passage of the egg by holding the egg from the outside of the bird (beneath the abdomen) and gently ease the egg outwards. In a bad case, the muscles are tight and will not allow the egg's passage. A calcium injection may help to relax the muscles; ask the vet. In a bad case, the egg may have to be broken and the bits extracted by forceps. This is best done at the vet's because any tearing will cause infection. Always give the bird a course of antibiotic to prevent infection of the oviduct. A bird with this condition may have no problem subsequently.

Egg Peritonitis

Egg peritonitis and associated abnormalities of egg-producing organs are the most common causes of death. The abdominal cavity can become inflamed due to the presence of egg yolk or even eggs with shells. A tear in the oviduct can result in eggs, particularly soft-shelled ones, passing into the abdomen, and forming a lump likely to cause the death of the duck. In a rare case, a duck might be saved by an operation conducted by a skilful veterinary surgeon (Renshaw, 1991).

Impaction of the Oviduct

The tube carrying the eggs from the ovaries to the cloaca may become impacted with cheesy material, and often broken egg shell, and may become infected.

Prolapse of the Oviduct

The lower part of the oviduct may protrude from the vent as a result of egg-binding or straining. This can gently be replaced in some cases. However, it is likely to be a persistent problem and the bird is best put down if the prolapse is bad or if the problem recurs.

Birds suffering from acute oviduct problems in the spring can die quite quickly for no obvious external reason. In chronic cases death may be delayed for several days or even weeks. The abdomen may become swollen and when it rests, the bird sits and bobs its tail. Birds obviously suffering from oviduct problems are best put down.

Avoid problems by ensuring that the ducks are not

Fig 185 Prolapse in a female Pekin during the laying season. The lower part of the oviduct and the cloaca is turned outwards. Because of the passage of eggs, this condition is virtually impossible to cure, and the bird is best put down. Birds in this condition must be removed from drakes, which can rip the exposed skin with their claws when mating.

too fat. Feed rations with calcium and vitamin D, lime the pasture if the pH is low and keep an eye on the birds in cold, wet weather in spring. Always provide mixed poultry grit. House laying ducks at night, if possible, to keep them warmer if there are problems with more than one bird. Products such as Calcivet sold by The Bird Care Company and Interhatch may be useful if on hand. Calcivet is a liquid calcium supplement, with vitamin D and magnesium.

Prolapse

In the female, this means that the lower part of the oviduct at the cloaca is turned outwards by the pressure of passing eggs (*see* above).

In the male, it means that the penis is carried outside the body and cannot be retracted. Drakes in this condition are no use for breeding. If they are pets or birds for exhibition only, the penis can be removed by a vet and the bird is otherwise all right. However, birds in this condition can be left alone for a week, or even longer, to see if they recover. They should be left on a pond to keep themselves clean. Sometimes the penis is fully retracted. If it is not quite retracted, the end becomes necrotic and falls off. Birds like this can still breed as the penis merely directs the semen. Keep an eye on affected birds until they seem fully recovered to make sure they do not develop a general infection.

Respiratory Infection

Lung congestion is rare in adult ducks kept in good conditions but occasionally strikes birds living outside, in periods of continuously wet, cold weather. An affected bird has difficulty in breathing and the lungs seem to rattle. The bird sits hunched up, and bobs its tail up and down to assist its laboured breathing. Prompt treatment with antibiotic such as Tylan 200 prescribed by the vet can save birds, but small Calls are more likely to succumb to the disease than larger ducks. Give a good course of antibiotic to make sure that it is effective; consult a vet on the time needed for effective treatment. Aspergillosis also affects the lungs (*see* above).

Sinus Infection

This is more common in ducks than geese. A swelling develops at the base of the bill so that the birds 'cheek' swells out. In a serious case, mucous may also been exuded from the nostrils. Treat with an appropriate antibiotic such as Tylan 200, prescribed by the vet (injection or soluble powder). Baytril is also used as a treatment. The rather hard swelling should soften as it responds to the antibiotic and, after about a week, the swelling should have gone. Sometimes the swelling remains despite the infective agent being dealt with. Make sure that the bird has lots of clean water to wash itself and the swelling should finally clear. Watch the bird to check that the infection does not re-occur in susceptible individuals, usually Call Ducks. Young birds can grow out of this condition as they get older.

Staphylococcus infections

These occur as localized infections of the tendon sheaths in the legs, and in the feet. The affected area is hot to the touch and swells; the bird limps. Infection can follow as a result of sprain injury or skin damage. Treat the bird with antibiotic injections. Infection is less likely in adults than in young birds. Avoid problems by having clean ground and feeding the birds well.

Fig 186 Buff Orpington with a sinus infection, which was cured by antibiotic treatment and better access to clean water.

Vitamin Deficiencies

These are unlikely to develop where birds are given some pelleted foods and wheat, and kept in natural conditions, i.e. on clean pasture. As with humans, a mixed diet including fresh food is best. Green foods supply vitamin A and K; sunlight allows the bird to utilize vitamin D; B vitamins and vitamin E are provided by cereal grains. Fishmeal and oils, added to starter and grower pellets, are especially important sources of vitamins. Deficiencies are most likely to occur in intensively reared ducks kept indoors and reared on starter rations designed for chickens.

Leg weakness is the most likely manifestation of vitamin deficiency. It may show as weak hocks in four-week-olds. This condition is best avoided. Adding niacin or brewers' yeast to the diet prevents the condition, or simply getting the ducklings out into grass runs. Sunlight also allows the birds to utilize vitamin D.

A diet of exclusively hen layers' pellets may cause problems for ducks, including poorly oiled, wet plumage. Ducks kept in these conditions improved when wheat was added to the diet, probably adding vitamin B which was deficient in the pellets.

Wet Feather

Ducks suffer from this condition more often than geese. The preen gland does not appear to work properly, so the bird cannot oil its feathers. The bird avoids water, because it will get wet and cold, so the feathers get in even worse condition. One cause may be a poor diet *(see* the paragraph above). Improve the amount of green food and vitamin B. Find a pellet ration specially for ducks. Add a small amount of cod liver oil to the pellets.

Treatment of the bird with Ivermectin (consult a vet for dosage), combined with the product Hexocil, removes parasites. Hexocil is shampoo which should be used as a weak solution, left on for 10 minutes, and then rinsed off very thoroughly. Wet feather is probably exacerbated by parasites, perhaps making the bird run down. Lice are said to cause excess preening thus damaging the feathers and causing wet feather. The problem may also be inherited, so if it is not cured by a better diet or environment, do not use the bird for breeding. The condition seems to be most common in coloured drakes such as Appleyards and Fawn Runners.

Wing Problems

Oar Wing/Angel Wing/slipped wing

This occurs between five and seven weeks when the blood-filled quills are sometimes too heavy to be supported by the final wing joint. It may happen in some seasons and not others, in ducklings bred from the same breeding pair. Its occurrence is therefore more often controlled by environmental factors than genetic disposition. It is particularly prevalent in the largest, fastest growers, i.e. the birds that would otherwise be the best specimens. *See* the section on feeding ducklings and correcting wings; treatment must be immediate to stop permanent deformity.

Dropped Wing

The duckling, between five and even ten weeks, allows its wing to fall below the normal position, but the quills do not rotate outwards. The muscles merely seem too weak to hold up the wing, which the duckling frequently hitches up. This often corrects itself, but can also be helped by taping the wing.

Rough Wing

The primary feathers do not fold snugly under the

secondaries. This is a result of high-protein feeding and too rapid growth. Regulate the food supply to prevent this condition.

Worms

These include the following.

Acuaria (renamed *Synhimantus spiralis*)
Humphreys (1975) refers to this particular roundworm being a problem in waterfowl. The worm damages the proventriculus so that the organ ceases to work and the bird gets thin and dies. This worm has a stage in its life-cycle when it lives in the body of the water flea *Daphnia*, which is common in ponds. A good strong flow of water is recommended as it stops the *Daphnia* from multiplying.

Amidostomum anseris
The gizzard worm of geese. It is said that this gizzard worm does not affect ducks. However, Urquhart *et al.* refer to *A. anseris* as being capable of causing 'heavy mortality in goslings, ducklings and other young aquatic fowls. The adult worms, bright red in colour and up to 2.5cm in length, are easily recognized at necropsy where they predominate in the horny lining of the gizzard.' If domestic ducks are affected they are less likely to pick up a high infestation because the majority of their food is from pellets and wheat. This is unlike geese, which graze much more, and therefore may pick up a high parasite load from the pasture.

Caecal worms (Hetarakis gallinarum)
Generally harmless unless in large numbers (*Poultry World Disease Directory*, 1994).

Gape Worms
Worms in the trachea (*Syngamus trachea* in poultry) cause birds to gasp for breath. This is less of a problem in ducks than in chickens, but it can make them cough. The birds otherwise behave normally and do not seem ill. Humphreys (1975) says that the eggs of the gapeworm *Cyathostoma bronchialis* can be picked up directly from the grass. Treat using a wormer such as levamisole or Flubendazole prescribed by the vet.

Worming Ducks

Ducks are best wormed by dosing themselves either via their food or their water. The mouth and throat of the smaller breeds such as Calls is too small to dose them individually with a drench unless a tube, operating the dosage by a hand pump, is inserted the correct distance down the throat. There is the danger of getting fluid into the windpipe and the lungs. Larger breeds such as Runners can be successfully individually treated in this way but it is often easier to treat the group in the drinking water or food.

Flubenvet (flubendazole) is the preferred vermifuge for birds, but is not licensed for ducks. Please consult your vet. It is administered as a powder, which adheres well to food. This is less suitable for geese, which will not always eat pellets. Flubendazole is usually administered over a week and therefore cannot be given at the point of sale. However, worms are not generally a problem in ducks. The commonest problem is coughing from throat worms at about twelve weeks, when the ducklets should be treated.

The dose for levamisole 7.5 per cent solution (out of a vets' handbook) for cage birds is 0.13–0.26ml/kg. The dosages multiplied up for ducks are given below, but the upper limits should not be scaled-up further. Larger animals have the dosage scales *down* per kilo, e.g. the dosage for cattle is 1ml/10kg.

The best way to give this wormer to ducks is in drinking water over a period of about half a day. Calculate the weight of the ducks and allow 0.5ml for the average ducks, and up to 1ml for the heavy ducks (some of the wormer will be wasted). The container has to be securely anchored, fairly small so the ducks do not use it as washing water to throw over themselves, and the only drinking water available. It must also be completely cleared up to make sure that they get the correct dose. There should be no need to worm ducklings before they are eight weeks old, and only do it then if there is a problem with coughing with throat worms. If birds are parasite-free, however, they will achieve optimal growth.

Flubendazole is the most effective wormer as it kills all worms, including tape worm.

Weight of duck (kg)	Weight (lb)	Lower and upper dosage (ml)	Average dosage
1	2.2	0.13–0.26	0.20
2	4.4	0.26–0.52	0.39
3	6.6	0.39–0.77	0.58
4	8.8	0.52–1.04	0.77

NB If delivered as a single dose down the throat with a syringe (no needle), the dosage should be diluted with 2ml of water.

Table to show the effectiveness of various wormers on some parasites

Wormer	***Amidostomum* (gizzard worm)**	***Syngamus* (gape worm)**	***Heterakis* (caecal worm)**
Fenbendazole found in *Panacur*	no; eggs not killed	yes	yes
Levamisole found in *Nilverm* and *Levacide*, 7.5 per cent	yes	yes	yes
Flubendazole	yes	yes	yes
Piperazine	no	no	partly

Ivermectin has been used for waterfowl in recent years as a systemic treatment for internal and external parasites. It is not licensed for poultry. It may be administered in tiny amounts on the skin, or injected. Ask you vet for details because the product comes in different forms. Some waterfowl, especially Toulouse geese and wildfowl, are sensitive to low doses and can die.

Few worming preparations have specifically been tested on ducks. Consult a vet before using them. See also V. Roberts, *Diseases of Free Range Poultry*.

Avian Influenza (Bird Flu)

Since this book was first published in 2001, avian influenza has become an issue in the UK. Several types of bird flu virus exist naturally in the wild bird population. Such viruses are of a low pathogenic type. These viruses can also affect domesticated birds. The intensification and scaling up of the poultry industry world wide, particularly in the Far East, has resulted in many opportunities for bird flu virus transmission, replication and mutation. Since 1996, the virus known as H5N1 has become highly pathogenic and has caused epidemics in commercial poultry in Hong Kong and Vietnam. The disease has now been limited in its spread in these areas by vaccination of poultry and enhanced biosecurity. Highly pathogenic H5N1 has also spread to Africa and Europe. Enhanced measures have been put in place in Europe to combat the spread of the disease. In the UK, poultry keepers with more than fifty birds are required to register with Defra. This ensures rapid communication in the event of outbreak, and allows for effective surveillance and containment. Unlike foot and mouth disease, the virus only moves on objects (including birds). Vaccination has not been used on a large scale in Europe and, at present (2008), is not authorized in the UK by Defra. See the Department for the Environment and Rural Affairs website at www.defra.gov.uk for good practice in biosecurity measures.

Useful Addresses

Allen & Page
Norfolk Mill, Shipdham, Thetford, Norfolk IP25 7SD
Tel: 01362 822900
e-mail sales@allenandpage.com
www.allenandpage.com

Ascott Smallholding Supplies Ltd (the mail order specialists)
The Old Creamery, Four Crosses, Llanymynech, Powys, SY22 6LP
e-mail sales@ascott.biz
www.ascott.biz

Avicultura (Dutch magazine, now on-line, mainly about birds and small animals)
www.avicultura.net

Aviornis UK
Laurie Crampton, Cold Arbor, Tytherington Lane, Bollington, Cheshire SK10 5AA
Tel/fax: 01625 573287

Barrier Biotech Ltd (disinfection and parasite control)
36 Haverscroft Industrial Estate, Attleborough, Norfolk NR17 1 YE
Tel: 01953 456363
www.barrier-biotech.com

Belgian Society of Domestic Waterfowl
Sonja Sauwens-Lambrighs, Grote Negenbundersstraat 42, 3511 Kuzingen-Hasselt, Belgium
Tel: 0032 112 54445

Blackbrook Zoological Park
BWA Conservation and Education Centre, Winkhill, near Leek, Staffordshire
Tel: 01538308293
www.blackbrookzoo.co.uk

The Bird Care Company (many bird care products)
Unit 9, Spring Mill Industrial Estate, Avening Road, Nailsworth, Glos GL6 OBU
Tel: 0845 130 8600 (local rates from UK landlines)
www.birdcareco.com

Bird Stevens and Co Ltd (rearing equipment, available at good smallholder suppliers)
Sun Works, Sun Street, Quarry Bank, Brierly Hill, West Midlands DY5 2JE
Tel: 01384 567381
www.birdstevens.co.uk

BOCM Pauls Ltd (duck food)
PO Box 2, Olympia Mills, Barlby Road, Selby, Yorkshire YO8 5AF
Tel: 01757 244000
www.bocmpauls.co.uk/bocmpauls/marsdens/homepage.jhtml

Brinsea Incubators Ltd
Station Road, Sandford, Avon BS19 5RS
Tel: 07979 305441
www.brinsea.co.uk

The British Waterfowl Association
Secretary: Mrs S Schubert, PO Box 163, Oxted, RH8 OWP
Tel: 01892740212
www.waterfowl.org.uk

Call Duck Association
Secretary: Graham Barnard,
Tel: 01558650532
www.callducks.net

Country Smallholding Magazine
Fair Oak Close, Exeter Business Park, Clyst Honiton, Devon, EX5 2UL
Tel: 01392 888475
e-mail info@countrysmallholding.com
www.countrysmallholding.com

Curfew Incubators
Curfew House, 4103 Route De Vintimille,
06540 Breil sur Roya, France

Fancy Fowl,
The Publishing House, Station Road,
Framlingham, Suffolk, IP13 9EE
Tel: 01728 622030

Feathered World
R. Batty, Winckley Press, 3 Winckley Street,
Preston PR1 2AA
Tel: 01772 250246

Gardencraft (poultry housing)
Tremadog, Porthmadog, Gwynedd LL49 9RC
Tel: 01766513036
Fax: 01766514364
www.gcraft.co.uk

German Domestic Duck Association
Hermann Lenz, Bergstrasse 6,
D-74867 Neunkirchen, Germany

Humane Slaughter Association
The Old School, Brewhouse Hill,
Wheathampstead, Herts AL4 8AN
Tel: 01582831919
Fax: 01582831414
e-mail: info@has.org.uk
www.hsa.org.uk

Indian Runner Duck Association
Secretary: Christine Ashton
Tel: 01938 554011
Chairman Julian Burrell,
Tel: 01579 340557
www.runnerduck.net

Interhatch, Incubator Specialists
Whittington Way, Old Whittington,
Chesterfield S41 9AG
Tel: 01246264646

MAFF Publications
London SE99 7TP
Tel: 08459 556000
www.defra.gov.uk/corporate/publications/ordering.htm

Marriage's Traditional Organic Feeds
W. H. Marriage and Sons Ltd, Chelmer Mills,
New Street, Chelmsford, Essex CM1 1PN
Tel: 01245 612000

Maywick Limited
Unit 7, Hawk Hill, Battlesbridge, Essex, SS11 7RJ
Tel: 01268573165
Fax: 01268573085

Poultry Club of Great Britain
www.poultryclub.org

Soil Association
www.soilassociation.org
Tel: 0117 314 5000

*Smallholder Magazine (*Began in about 1910 for
small farmers, allotment holders and gardeners)
Editor: Liz Wright
Tel:01354 741538
www.smallholder.co.uk

Victoria Waterfowl Association
Dr H. M. Russell, 22, Georges Road, The Patch,
Victoria 3792, Australia

Bibliography

Ambrose, A., *The Aylesbury Duck* (Buckinghamshire County Museum, 1991)

American Poultry Association, *American Standard of Perfection* (1905)

Appleyard, R., 'The Indian Runner Duck. – I', *The Feathered World* (18 September 1925)

Appleyard, R., 'The Indian Runner Duck. – II', *The Feathered World* (25 September 1925)

Anderson Brown, Dr. A.F.Anderson, *The Incubation book* (Wheaton, 1979)

Appleyard, R., 'The Indian Runner Duck', *Feathered World Yearbook*

Appleyard, R., in *Poultry World* (26 January 1934)

Appleyard, R., *Ducks: Breeding Rearing, and Management* (Poultry World Ltd, 1937)

Ashton, Ashton and Stanway, in *Fancy Fowl*, (1998)

British Poultry Standards 4th edition (1954)

British Poultry Standards 5th edition, Roberts, V. (ed.) (1997)

British Waterfowl Standards, reprinted from *British Poultry Standards*, 4th edition (Butterworth, 1982)

British Waterfowl Standards, Ashton & Ashton (ed.) (BWA, 1999)

Broekman, K., in *Waterfowl (*spring 1987)

Brooke, M., & Birkhead, T., *The Cambridge Encyclopedia of Ornithology* (CUP, 1991)

Brown, E., *The Duck Breeding Industry of England and America* (Reliable Poultry Journal Publishing Company, 1910)

Brown, E., *Poultry Breeding and Production* (Caxton, 1929)

Brown, E., *Races of Domestic Poultry*

Browne D. J., *The American Poultry Yard* (1853)

Buhle, 1860, cited in Schmidt (1989)

Burdett, W., 'Crown Roots', *Waterfowl Yearbook* (BWA, 1977–78)

Charnock Bradley, O., revised by T. Grahame, *The Structure of the Fowl* (Oliver and Boyd, 1950)

Cook, W., *Ducks: and How to Make them Pay* (circa 1899)

Coutts, J.A., in *The Feathered World* (1923)

Coutts, J.A., 'Crested Runners', *The Feathered World* (4 December 1925)

Coutts, J.A., as 'Red Feather', *The Feathered World*, p. 442 (1925)

Curtis, P., *A Handbook of Poultry and Game Bird Diseases* (Liverpool University Press, 1996)

Davidson and Chisholm, in *The Feathered World* (1929)

Dixon, E.S., *Ornamental Domestic Poultry* (1848)

Donald, J., *The Indian Runner Duck: its History and Description* (c.1890)

Donald, J., letter to *The Feathered World*, p. 1,131 (1905)

Durigen, B., cited in Schmidt (1989)

Earle, R. in *Fancy Fowl* (November 1998)

Ellis, W., *The Country Housewife's Family Companion* (1750)

Emmet, T., a reader's letter published in *Fancy Fowl* (July 1999)

Freethy, R., *How birds work* (Blandford Press, 1982)

Grow, O., *Modern Waterfowl Management* (American Bantam Association, 1972)

Hall, Sherwin A. 'Duck Plague in Britain' *Waterfowl Yearbook* (1973–74)

Hams, F., *Domestic Ducks and Geese* (Shire Publications, 2000)

Holderread, D., *Breed Bulletin: French Rouen Ducks* (1985)

Horton, L., in *Avicultura U.K.* (1995, Vol. 1, No. 3)

Humane Slaughter Association, *Practical Slaughter of Poultry*

Humphreys, P.N., 'Worms in Waterfowl', *Waterfowl* (British Waterfowl Association, 1975)

Hurst, J.W., *Utility Ducks and Geese* (Constable and Co.,1919)

Ives, P., *Domestic Geese and Ducks* (Orange Judd, 1947)

Johnsgard,, P.A., *Ducks in the Wild* (Prentice Hall, 1992)

Jull, M.A., 'Fowls of Forest and Stream', *National Geographic Magazine* (March 1930)

Kear, J., 'Dutch Decoys: with Particular Reference to Bird Ringing', *Waterfowl Yearbook* (1993–94)

Laing, P.W., *The Poultry Farmer's and Manager's Veterinary Handbook* (The Crowood Press, 1999)
Long, R.A., 'The Khaki Campbell', *Ducks* (1926)
Lunz, R. , 'Kaleidoscope of colours – the Saxony Duck', *Gerflugel Borse* (July 1995)
Mattocks, J.G., 'Goose Feeding and Cellulose Digestion' , *Wildfowl 22* (The Wildfowl Trust, 1971)
Ministry of Agriculture, Fisheries and Food, *Codes of Recommendations for the Welfare of Livestock: Ducks* (PB 0079, 1999)
Moubray, B., *Treatise of Domestic Poultry* (1815)
Nolan, J.J., *Ornamental, Aquatic, and Domestic Fowl and Game Birds* (1850)
Owen, M., *Wildfowl of Europe* (Macmillan, 1977)
Pearce, F. H., 'So what if the power goes off?', *The Waterfowl* (Victorian Waterfowl Association, Australia, 1998)
Poultry Club Standard of Excellence (1865)
Poultry Club Standards, Threlford (ed.) (1901, 1905)
Poultry Club Standards, Broomhead, W.W. (ed.) (1910, 1922, 1923, 1926, 1930)
Poultry World Disease Directory (1994)
Poultry Yearbook (1925)
Powell, J.C. in *Waterfowl*, (British Waterfowl Association, spring 1993)
Powell-Owen, W., *Duck-keeping On Money-Making Lines* (George Newnes Ltd., 1918)
Powell-Owen, W., *The Complete Poultry Book* (Cassell and Co. Ltd, 1953)
Rankin J., *Profitable Pekin Ducks* (Reliable Poultry Journal Publishing Company, 1910)
Renshaw, J., 'Egg-binding – with a Difference', *Waterfowl Yearbook* (British Waterfowl Association, 1991–92)
Roberts, M. and V., *Domestic Duck and Geese in Colour* (Domestic Fowl Trust, 1986)
Roberts, V., *Modern Vermin Control* (1989)
Roberts, V., *Diseases of Free Range Poultry* (Whittet Books, 2000)
Schippers, H., in *Fancy Fowl* (June 1997)
Schmidt, H., *Puten, Perlhuhner, Ganse, Enten* (Neumann-Neudamm,1989)
Scott, P., *A Coloured Key to Waterfowl of the World* (W.R.Royle & Son, 1972)
Selten, 'Pekin Ducks', *Avicultura International* (1996, Vol. 2 Nos 1 and 2)
Sewell, F., 'Ducks and geese', *Reliable Poultry Journal* (1910)
Sheraw, C. D., *Successful Duck and Goose Raising* (Stromberg Publishing Company, 1975)
Sheraw, C.D., *The Call Duck Breed Book* (American Bantam Association, 1983)
Sheraw, C. D., *The East Indie Duck* (American Bantam Association, 1990)
Spinke, S., 'The Rouen's Wondrous Raiment', *The Feathered World* (30 March 1928)
Stapel, C., in *Waterfowl* (British Waterfowl Association, 1976)
Stavelely, C. R., 'Hook-Billed Ducks', *The Feathered World* (1913)
Stevens, L., 'Poultry genetics', *Fancy Fowl* (September 1995)
Stromberg, L., *Poultry of the World* (Silvio Mattachione, 1996)
Taverner, B., *Great Maritime Routes* (Macmillan, 1972)
Taylor, E.A., *Runner Ducks* (Country Life Ltd., 1918)
Tegetmeir, *The Poultry Book* (1867)
Thear, K., 'Complete guide to chickens. Part 2:Feeding' , *Country Smallholding* (1998)
Thompson, J.M., 'Will they, won't they? Won't they join the dance?', *Waterfowl* (British Waterfowl Association, 1984)
Todd, F. S., *Waterfowl: Ducks, Geese and Swans of the World* (Sea World Press, 1979)
Urquart, G.M., Armour, J., Duncan, J.L., Dunn A. M., Jennings, F.W., *Veterinary Parasitology* (Blackwell Science)
van Gink, 'The New French Rouen Duck', *The Feathered World* (27 November 1931)
van Gink, 'The Dutch Call Ducks', *The Feathered World* (1932)
Walton, J. W., 'The Indian Runner Duck Club and its Standard', *The Feathered World*, p. 47 (1908)
Weir, H., *Our Poultry* (Hutchinson and Co., 1902)
Willughby, F., *Ornithologie , Book III* (1678)
Wingfield, Rev. W. and Johnson, G.W., *The Poultry Book* (1853)
Wilson-Wilson, T., 'Indian Runner Ducks at Kendal Show', *Poultry* (13 November 1896)
Wright, Lewis, *The Illustrated Book of Poultry,* (Cassell and Co., 1874)
Wright, Lewis, *The New Book of Poultry* (1902, also 1910)

Index